Curbing Energy Poverty in Sub-Saharan Africa

Pitshou Ntambu Bokoro · Gulshan Sharma

Curbing Energy Poverty in Sub-Saharan Africa

A Multi-Objective Optimization Approach for Achieving Universal Access

Springer

Pitshou Ntambu Bokoro
Department of Electrical Engineering Technology
University of Johannesburg
Johannesburg, Gauteng, South Africa

Gulshan Sharma
Department of Electrical Engineering Technology
University of Johannesburg
Johannesburg, Gauteng, South Africa

ISBN 978-3-031-90071-6   ISBN 978-3-031-90072-3 (eBook)
https://doi.org/10.1007/978-3-031-90072-3

This Springer imprint is published by the registered company Springer Nature Switzerland AG
The registered company address is: Gewerbestrasse 11, 6330 Cham, Switzerland

Preface

Since the dawn of civilization, humanity's socio-economic progress has been intricately linked to its ability to access and effectively utilize energy. From the rudimentary energy applications of ancient times to its transformative role in driving industrialization, electrical energy has emerged as a fundamental necessity for growth and development. Yet, paradoxically, an estimated 775 million people — the majority of whom live in Sub-Saharan Africa (SSA) — continue to lack access to modern energy, depriving them of the essential foundation for socio-economic advancement and sustainable development. At the heart of energy poverty lies the persistent reliance on traditional fuels such as charcoal, paraffin, and woodstoves. This dependence not only hampers access to clean energy but also poses severe health risks. The World Health Organization (WHO) predicts that deaths related to woodstove usage could rise to 1.5 million globally by 2030, underscoring the urgent need for clean fuels (green molecules) and the imperative to preserve the planet's ecosystem.

In SSA, energy poverty is largely perceived as an issue of accessibility, stemming from the limited physical connection to energy grid systems. Consequently, strategies aimed at addressing this challenge focus on leveraging renewable energy-based decentralized systems, implementing progressive regulatory policy reforms, and, most critically, developing innovative funding models to ensure sustainable solutions and energy justice. Despite advancements in renewable energy technologies and the adoption of market-based policies across SSA, progress in expanding access to energy services has remained limited, highlighting the shortcomings and inefficiencies of the models implemented to date.

This book underscores the importance of a multidimensional approach to addressing energy poverty, advocating for a comprehensive strategy that integrates three interrelated dimensions: availability, accessibility, and affordability. It introduces a Multi-Criteria Decision-Making (MCDM) framework built on a dynamic model linking off-grid energy supply with rural households engaged in crop farming as a key economic activity. The framework employs the Non-Dominated Sorting Genetic Algorithm II (NDSGA II) — a multi-objective optimization strategy — to balance the trade-offs between costs and revenues in ensuring sustainable delivery of energy services to rural households. By affirming the interconnected trilogy of availability, accessibility, and affordability, the approach highlights that while costs are primarily tied to ensuring energy availability, revenues are pivotal in enhancing both accessibility and affordability. This holistic strategy effectively addresses the complex nature of energy poverty in rural areas.

Johannesburg, South Africa
December 2024

Pitshou Ntambu Bokoro
Gulshan Sharma

Competing Interests The authors have no competing interests to declare that are relevant to the content of this manuscript.

Contents

Acronyms

AFCP	After Fuel-Cost Poverty
AFD	Agence Française de Développement
AfDB	African Development Bank
ANSER	Agence Nationale de l'Electrification et des Services Energétiques en milieux rural et périurbain
APP	African Power Pool
ARE	Agence de Régulation de l' Electricité
BIPV	Building Integrated Photovoltaic
BOT	Build Operate and Transfer
CAGR	Compound Annual Growth Rate
CAPP	Central African Power Pool
CDA	Community Development Association
CEEPI	Composite Electrical Energy Poverty Index
CEPI	Comprehensive Energy Poverty Index
COP	Conference of Parties
COPD	Chronic Obstructive Pulmonary Disease
CSP	Corporate Social Responsibility
CVM	Contingent Valuation Method
DISCo	Distribution Companies
EAPP	East African Power Pool
ECCAS	Economic Community of Central African States
ECDPM	European Centre for Development Policy Management
ECOWAS	Economic Community of West African States
EDC	Electricité du Congo
EDG	Electricity Distribution of Ghana
EDI	Energy Development Index
EEP	Electrical Energy Poverty
ENK	Energie du Nord Kivu
EPOV	Energy Poverty Observatory
ERA	Electricity Regulation Amendment
ERC	Energy Regulatory Commission

ESCOM	Electricity Supply Company of Malawi
ESKOM	South African Power Utility
ESMAP	Energy Sector Management Assistance Programme
ESS	Energy Storage Systems
ETP	Energy Transition Plan
EU	European Union
FIT	Feed-in Tariff
GDP	Gross Domestic Product
GEDAP	Ghana Energy Development and Access Project
GEF	Global Environment Facility
GENCo	Generating Companies
GFDRR	Global Facility for Disaster Reduction and Recovery
GHG	Green House Gas
GRIDCo	Ghana Power Utility
HDI	Human Development Index
HEAI	Household Energy Access Index
HPP	Hydropower Plant
IAEA	International Atomic Energy Agency
ICA	International Compliance Association
IEA	International Energy Agency
IEP	Integrated Energy Plan
ILO	International Labour Organization
IMF	International Monetary Funds
IPP	Independent Power Producer
IRENA	International Renewable Energy Agency
IRP	Integrated Resource Plan
KCM	Kamoto Copper Company
KPLC	Kenya Power and Lighting Company
LCOE	Levelized Cost of Energy
LTS	Long-Term Strategies
MCDM	Multi-Criteria Decision-Making
MEC	Manila Electric Company
MEPI	Multi-Dimensional Energy Poverty Index
MIS	Minimum Income Standard
MOO	Multi-Objective Optimization
MTF	Multi-Tier Framework
NCQG	New Collective Quantified Goal
NDP	National Development Plan
NDSGA	Non-Dominated Sorting Genetic Algorithm
NEP	National Energy Policy
NERC	Nigerian Electricity Regulatory Commission
NERSA	National Energy Regulator of South Africa
NOAA	National Oceanic and Atmospheric Administration
NPV	Net Present Value
NREAP	National Renewable Energy Action Plan

NREEEP	National Renewable Energy and Energy Efficiency Policy
OECD	Organisation for Economic Co-operation and Development
PAYG	Pay as You Go
PLC	Programmable Logic Controller
PLS SEM	Partial Least Squares Structural Equation Modelling
PPA	Power Purchase Agreement
PPP	Public Private Partnership
PSE	Poverty and Social Exclusion
PV	Photovoltaic
RE	Renewable Energy
REIPPPP	Renewable Energy Independent Power Producer Procurement Programme
REIS	Renewable Integrated Energy System
RFP	Request For Proposal
SADC	Southern African Development Community
SAPP	South African Power Pool
SDG	Sustainable Development Goals
SHS	Solar Home Systems
SNEL	Societé Nationale de l' Electricité
SOU	State Owned Utilities
SSA	Sub-Sahara Africa
TCN	Transmission Company of Nigeria
TOU	Time of Use
TSO	Transmission System Operator
UK	United Kingdom
UN	United Nations
UNDESA	United Nations Department of Economic and Social Affairs
UNDP	United Nations Development Programme
UNEP	United Nations Environment Programme
UNFCCC	United Nations Framework Convention on Climate Change
UNIDO	United Nations Industrial Development Organization
UNSDS	United States Summit on Sustainable Development
US AID	Agency for International Development
US EIA	United States Energy Information Administration
US	United States
USD	United States Dollar
WAEPI	Weighted Average Energy Poverty Index
WAPP	Western African Power Pool
WHO	World Health Organisation
WTP	Willingness to Pay
ZESA	Zimbabwe Electricity Supply Authority

Chapter 1
Energy and Human Race Evolution

1.1 Introduction

History of energy utilization by the human race is traced back in the period before the Middle Ages (Cleveland 2019; Grinin et al. 2015; Sorensen 2013). Despite the inefficiency of the form of energy consumed at the time of great civilisations of antiquity, the usage of energy was purely motivated by basic needs for survival (cooking, heating, lighting, etc.). This could be referred to as rudimentary or the early era of industrialization (Crafts 1985; Mokyr 1999).

However, hard labour endured to produce energy could be regarded as the main driver for the quest for efficient ways of energy production. This has led to continued evolution of the human race in terms of industrialization, which is very often characterised by the emergence or discovery of a new and dependable source of energy to fuel major or dominant socio-economic activities (Landes 2003; Morris 2010). In this respect, energy has been found to be at the centre of all industrial revolutions the human race has ever known to date (Smil 2017; Herring 2010). Accordingly, the first industrial revolution (1760–1840) was fundamentally epitomised by the introduction of coal-powered steam engines, while the second industrial revolution (1870–1914) witnessed the emergence of magnetism and electricity, which enabled mass production in manufacturing plants. Although the third industrial revolution (1969–2005) is often associated to the emergence of digital electronics, computer systems, programmable logic controllers ($PLCs$) giving rise to production automation (Bressler 2001; Brynjolfsson and McAfee 2011), it is however worth noting that the introduction of nuclear energy occurred during this time (Sovacool 2011). The advent of the fourth industrial revolution (2007) has enabled intelligent usage of several types of energy sources (alternative forms of energy and fossil) in combinational and/or sequential mode of operation in a bid to reduce dependency on fossil fuels owing to greenhouse gas emissions (GHGs), climate change and global warming.

In addition to the emergence of a new source of energy during each of the industrial revolutions, there is also a number of dominant socio-economic activities, which

P. N. Bokoro and G. Sharma, *Curbing Energy Poverty in Sub-Saharan Africa*,
https://doi.org/10.1007/978-3-031-90072-3_1

served the purpose for which the new source of energy was required. Owing to the challenges of climate change and global warming, which pose significant threats to the planet and hence to the existence of the human race (Hsiang et al. 2017; Hansen et al. 2012), the question of sustainable energy has become central and therefore quite legitimate to the survival and continued evolution of the human race. However, the attainment of sustainable energy cannot be achieved overnight. It does require appropriate and specific actions (investments, measures and policies) to be formulated and implemented for a certain period of time deemed necessary to significantly reduce, or if possible, to subvert the dependency on fossil energy.

Therefore, energy transition is critical and unavoidable process towards decarbonation or the carbon-free or green energy economy. Fundamentally, this transition is intended to significantly slow-down or reduce carbon emissions or footprints by adopting clean and efficient technologies of energy production and usage (UNDP 2020a; UNIDO 2021). The United Nations (UN) have developed a set of comprehensive goals and targets, known as Sustainable Development Goals ($SDGs$). These objectives have been adopted by almost all the nations to the extent of being incorporated into many government's programmes (Sachs etal. 2019). Furthermore, the emphasis on access to low-carbon and affordable energy is of paramount importance. Despite this recommendation, some parts of the world (Sub-Sahara Africa, South Asia, etc.) and millions of people continue to experience lack of access to energy. These populations with no access to energy are subjected to dehumanised living conditions or abject poverty.

1.2 Ending Three Centuries of Fossil Energy Usage

Over the last three centuries, fossil fuel-based energy (coal, oil, natural gas) has been the most dominant energy resource used to sustain economic expansion and development of nations around the globe (Gately 2015; Kallis et al. 2012). In principle, fossil fuel fundamentally consists of carbon and hydrogen as its main constituents (Mathur and Rao 2018). The combustion process of this fuel dissipates significant amount of heat, which is utilised to power engines and generate base load electricity through steam turbines (Wang et al. 2020). The estimated remaining number of years for the utilization of fossil fuel resources is given in Fig. 1.1.

This indicates that these resources will still be available for use for several decades to come. In addition, the wealth of several powerful nations as well as the income of many households are dependent on this economic sector. This established reality as well as the low cost of these hydrocarbon-content resources consist of major determining factors that support the continued reliance on fossil-fuel based energy despite environmental pollution and greenhouse effect associated to this source of energy (Chiu and Chu 2021, Schiffer et al. 2018). According to Ripple et al. (2021), the catastrophic effects of fossil energy usage, which started back in the first industrial revolution, have significantly progressed to the extent of imminent threat of irreversible destruction of the planet ecosystem and human life extinction.

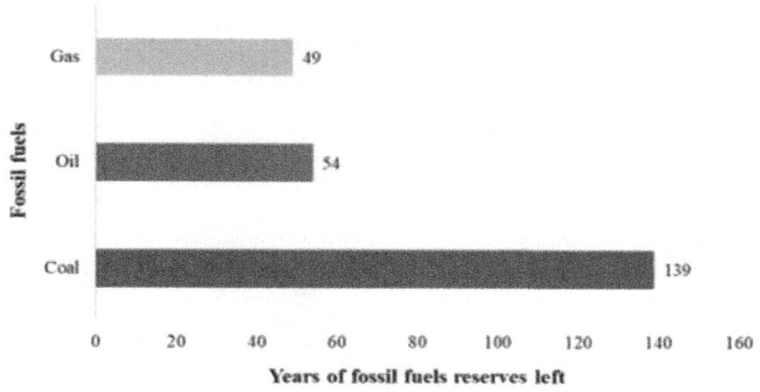

Fig. 1.1 The estimated remaining number of years for the utilization of fossil fuel resources

1.2.1 Global Warming and Climate Change

According to Fuller et al. (2020), environmental pollution consists of one of the serious threats to human life as it negatively impacts the ecosystem: air, water, and land. Manisalidis et al. (2020) refer to pollution as the introduction into the environment of higher concentrations of harmful or toxic substances to humans and other living organisms. The correlation between atmospheric pollution, GHG emissions (carbon dioxide, methane, nitrous oxide, fluorinated gases) and fossil fuel-based energy usage has been extensively discussed in the various literature sources (Jackson et al. 2020; Jones et al. 2023; EPA 2021). This suggests that the release of GHG emissions in the atmosphere mainly originates from the usage of fossil fuel for energy consumption (electricity production, heating and transportation). The various sources of GHG emissions are given in Fig. 1.2.

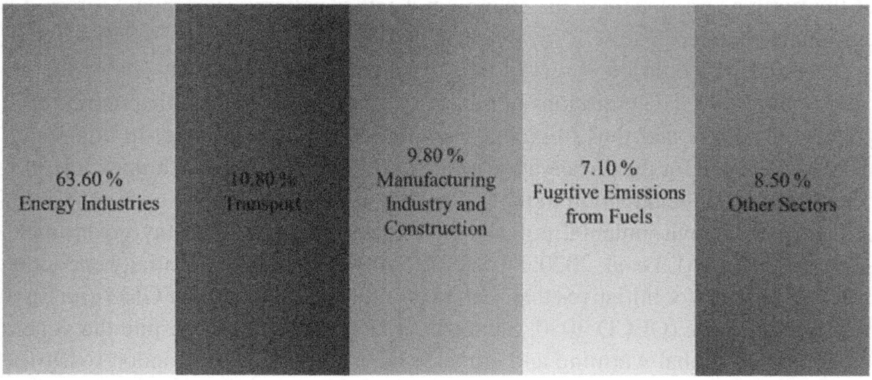

Fig. 1.2 Sources of GHG Emissions released in the atmosphere

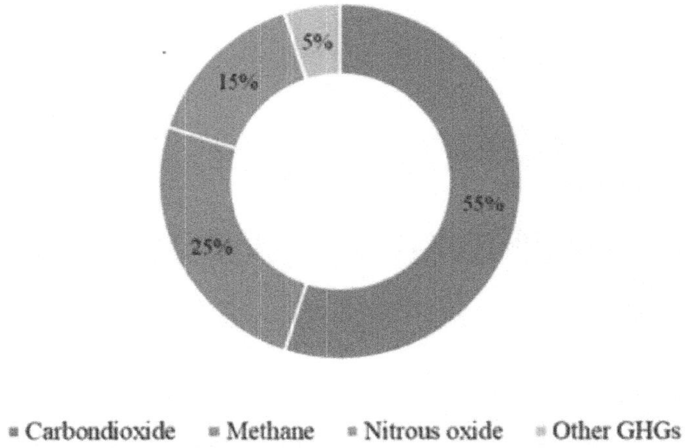

Fig. 1.3 Percentage contribution for each GHG in the atmosphere

In 2022, the US National Oceanic and Atmospheric Administration (NOAA) reported that carbon dioxide, which is the most prevalent GHG in the atmosphere, reached the highest ever concentration of 410 parts per million (ppm) since the begining of the industrial revolution (280 ppm) (NOAA 2020; Nilkar et al. 2023). This accumulated concentration of GHGs in the atmosphere is responsible for greenhouse effect. This refers to the process whereby radiations from the sun are absorbed by GHGs in the atmosphere and not reflected back into space. This insulates the surface of the Earth and prevents it from freezing (Held and Soden 2020). The percentage contribution of GHGs in the atmosphere is indicated in Fig. 1.3.

Among the major consequences of greenhouse effect features the progressive increase in the average temperature of the Earth's atmosphere, which is best known as global warming. However, the long-term effect of this increasing rise in temperature of the Earth's atmosphere is the change in weather patterns, commonly referred to as climate change.

Therefore, it could be observed that fossil fuel-based energy is one of the primary sources of GHG emissions in the atmosphere, and significantly contributes to greenhouse effect and thus to global warming and climate change. In this regard, increased efforts, for the preservation of the Earth's eco-system and human life, have been given absolute attention to the extent that major shift or re-alignment in terms of energy and environmental policies have been endorsed by many governments around the world (Lu et al. 2020; DESA 2015). These policies essentially encourage interventions and/or initiatives that seek to promote the reduction of GHG emissions in the atmosphere (OECD 2020; Santos et al. 2022). However, despite the general agreement on global warming and climate change, attempts to abandon fossil-fuel based energy sources or to implement drastic reduction of GHG emissions are still perceived to be slowing-down the current pace of industrialization, urbanization or socio-economic development of nations.

1.2.2 Green Energy Revolution

The reduction of GHG emissions should, at any costs, be balanced with the quest for industrialization and urbanization around the world in order to keep up with socio-economic growth and development. In this light, reliance on low-GHG emissions or clean technologies (Solar, wind, geothermal, hydro, nuclear, etc.) to generate electricity is currently regarded as the most sustainable approach to maintain the balance between socio-economic developmental needs and environmental pollution. This triggers the begining of the end of the three centuries of absolute dependence on fossil-fuel based energy, otherwise referred to as dirty energy, which favours the emergence of clean or green energy revolution. The main sources or technologies that lead to electricity generation with lower amounts of GHGs are given in Fig. 1.4.

As previously discussed, the correlation between energy and socio-economic development, such as consistently observed in the course of human race evolution, has led to strict regulatory frameworks (carbon tax, GHG emission penalties, carbon capture and storage, etc.) and massive investment incentives (renewable energy tax incentives, green energy investment, EU renewable energy financing mechanism, EU funding for offshore renewables, etc.) in carbon-neutrality or green energy projects

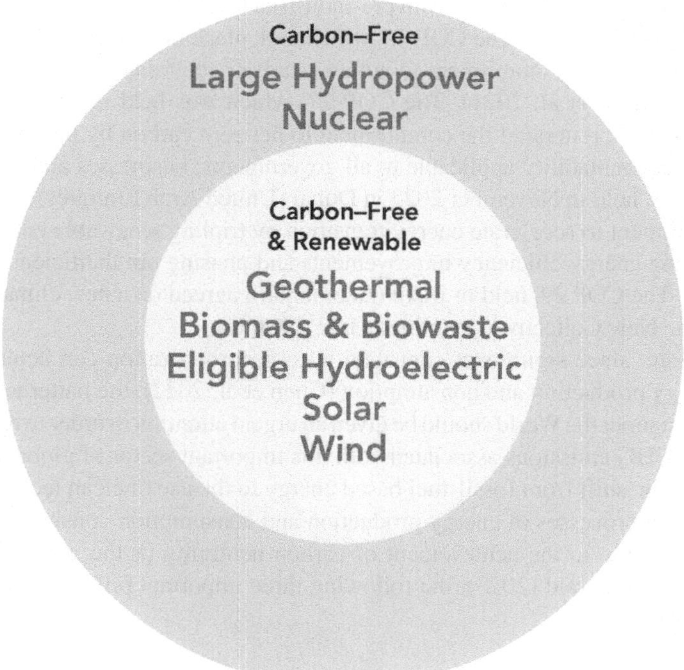

Fig. 1.4 Main sources or technologies that are used for electricity generation with lower amounts of GHGs

in effort to promote and achieve decarbonization of economy, and hence the adoption of the term: green economy (UNEP 2011; UNDP 2020b). Therefore, green energy or low-GHG energy sources are expected to become the dominant driver of socio-economic activities of the human race in the next few decades. However, due to the scarcity of water resources and the unavailability and/or negative perceptions towards nuclear technologies in several countries, it cannot be possible to instantly abandon the current reliance to dirty energy in order to quickly embrace the green energy revolution, which in this case, should be completely dependent on the stochastic nature of renewable resources (solar and wind). In recognition to these imperatives, timeframes related to key targets have been proposed in order to manage the compliance with the necessary interventions and measures to be implemented during the transition to green energy revolution.

1.3 Energy Transition

Owing to the devastating effects of global warming and climate change, an international agreement was signed in December 2015, at the 21st conference of parties (COP 21) in Paris, which set the target of limiting global warming by the end of this century to below 2 °C compared with pre-industrial levels and preferably limiting it to 1.5 °C (UNFCCC 2016). The COP 26, which took place in November 2021 in Glasgow, confirmed this commitment to achieve carbon neutrality by 2050 (UNFCCC 2016; Mountford et al. 2021). The COP 27, which was held in Sharm el-Sheikh (Egypt) in 2022, reiterated the commitment to net zero carbon by proposing mechanisms of accountability applicable to all governments, businesses and institutions. The COP 28, held in November 2023 in Dubai (United Arab Emirates), emphasized the commitment to accelerate energy transition by tripling renewable energy capacity, doubling energy efficiency improvements and phasing out inefficient fossil fuel subsidies. The COP 29, held in Baku (Azerbaijan), agreed on a new climate finance goal — the New Collective Quantified Goal (NCQG).

Therefore, since significant contribution to decarbonization can actually come from energy production and consumption (Chen et al. 2022), the patterns of energy usage throughout the World should be given an urgent attention in order to reduce high carbon or GHG emissions associated with this important sector of global economy. Therefore, the shift from fossil-fuel based energy to the use of clean technologies or non-emitting processes of energy production and consumption consists of essential step or transition to the achievement of carbon neutrality or the green revolution. According to Emerald (2023), the following three important pillars are assigned to the energy transition:

1.3.1 Green Electrons

This concept implies that the complete electricity generation is achieved using non-emitting or clean sources. This could also be produced from energy storage systems (ESS), which are designed to store energy for future supply capacity. Therefore, this ensures clean conveyance of this commodity through the grid, and ultimately to consumption centres whereby the various types of consumers may be able to tap into the grid.

1.3.2 Green Molecules

This concept refers to non-emitting fuels, which are meant to release their stored energy through clean processes. The lifecycle of the non-emitting fuels is environmental friendly with no or insignificantly low GHG emissions. Clean hydrogen and synthetic fuels could be counted among these types of fuels.

1.3.3 Green Bits

This refers to the digitisation of the energy system, which may consists of several applications: performance optimization of production assets in order to achieve efficient utilization of energy in buildings, vehicles and factories.

According to Arena's World Energy Transition Outlook charts (IRENA 2022) the alignment of the climate-safe future with the Paris Agreement dictates the conception of a roadmap, which accelerates the global energy transition by positioning electrification and efficiency as the main drivers of change, supported by renewables, hydrogen and sustainable biomass. This is indicated in Fig. 1.5.

It is worth noting that the 1.5 °C target set in the Paris Agreement requires capital investments of about 5.7 trillion USD per year until 2030. In order to subvert the risk of not being able to fundraise at that level, the IRENA World Transition Outlook (2022) suggests that the amount of 0.7 trillion USD, which consist of annual investments in fossil fuels, be re-invested into energy transition technologies. Therefore, additional capital could be sourced from the private sector. In addition, a comprehensive set of policies providing details of all possible technological solutions is required in order to achieve the necessary levels of deployment in 2030. In order to federate nations' support for a trigger to the just transition, a holistic global policy framework may be required to trigger international finance flows, capacities and technologies (IRENA 2022). This could be illustrated in Fig. 1.6.

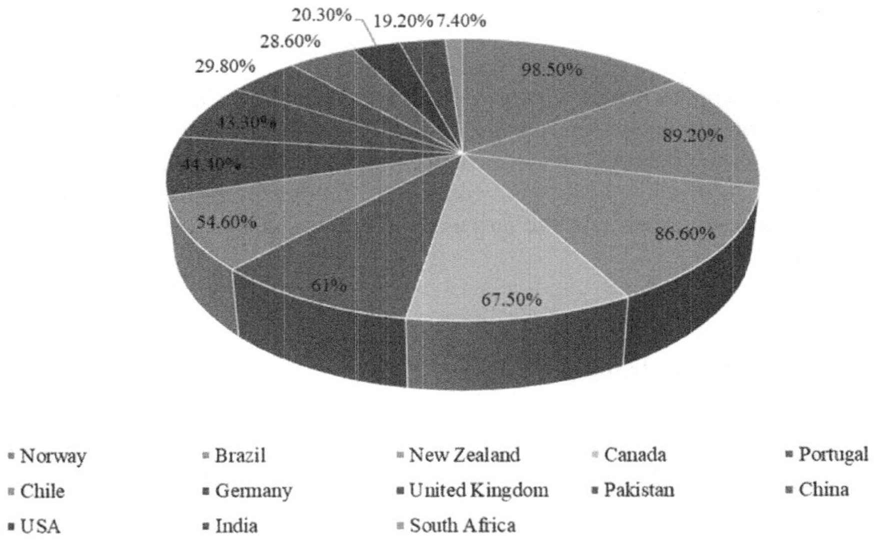

Fig. 1.5 Percentage usage of green electricity across the world

Fig. 1.6 Finance flows, capacities and technologies that trigger the just transition

1.3.4 Just Transition

The concept of just transition originated in the early 1970s when stricter environmental regulations in the US adversely impacted employment in industries such as oil, chemical, and atomic energy that failed to meet these standards. To protect workers facing mass unemployment, trade unions and allied groups advocated for a just transition, rather than calling for the repeal of environmental protection laws (Wang and Lo 2021). In recent years, this concept has resurfaced in the context of achieving climate goals, emphasizing the need to ensure that all communities, workers, and social groups are included in the transition to a net-zero future. According to the International Labour Organisation (ILO), the just transition implies greening the economy in a manner that is fair and inclusive for all, ensuring the creation of decent work opportunities and leaving no one behind (UNDP 2022). Therefore, a just transition not only promotes more ambitious environmental and climate action but also advances progress towards achieving the SDGs, particularly those focused on affordable and clean energy, decent work and economic growth, reduced inequalities, and responsible production and consumption. Recognising the importance of this concept, government worldwide are beginning to integrate just transition principles into their short and long-term climate plans, known as National Determined Contributions (NDCs) and Long-Term Strategies (LTS), respectively.

1.4 Universal Access to Clean and Affordable Energy

Responding to the key issues of global concern, which include among others: global warming and climate change, the UN Summit on Sustainable Development held in September 2015 — which had witnessed the participation and contribution of several heads of government and/or state, senior representatives of government, leaders of global business entities and civil society movements — formally adopted the 2030 Agenda for Sustainable Development with 17 SDGs (SDS 2015; UN 2018). The transformative, inclusiveness and universality nature of the adopted 2030 agenda was also emphasised under the motto "leaving no one behind". Therefore, the 2030 agenda proposes national ownership by governments as key for implementation, encourages the strengthening of international institutions with multilateral regional and international organisations that seek to promote sustainable development, as well as the alignment between national, regional and international policies and priorities. The list of SDGs proposed in the 2030 agenda is indicated in Fig. 1.7.

Among the SDGs, access to affordable and clean energy (Goal 7) is meant to address the challenges related to global warming and climate change by reducing GHG emissions in the energy production and consumption sector. The following three targets are associated with this goal:

Fig. 1.7 The United Nations' SDGs (UNSDS 2015)

- To ensure universal access to affordable, reliable, and modern energy services
- To increase the share of renewable energy in the global energy mix.
- To improve energy efficiency in all sectors of the economy.

In order to achieve these targets, it is recommended that governments, businesses, and civil society work together to develop policies and initiate investments that are meant to support the transition to clean and renewable energy sources (wind, solar, and hydropower, etc.) while promoting energy efficiency.

Despite these measures, millions of people around the World continue to experience energy deprivation: either in a form of inaccessibility of this service (Ritchie et al. 2022; UNDP 2020c) or the inability to afford the cost of energy services (Lapaz et al. 2022). These two concepts contribute to what has been identified in the literature since 1990 as energy poverty. Therefore, energy poverty, if not significantly reduced, will continue to slow down the progress related to the implementation of universal access to energy and sustainability. The following causes are generally associated with energy poverty (Castaño-Rosa et al. 2020; Ahmed and Gasparatos 2020; Smail 2023):

- Low income
- Energy inefficient dwellings
- High cost of energy
- Lack of energy carriers or grids
- Households' needs
- Lack of policy recognition

Therefore, the quest for universal access to energy services imposes the need for curbing the major causes responsible for slow progress towards this goal by implementing policies and initiatives that are likely to favour investment in energy infrastructure, support innovative business models, and address socio-economic barriers to energy access.

1.5 Energy Poverty

1.5.1 Contextual Background

Although a universally accepted definition of energy poverty does not exist, two fundamental concepts are commonly employed to describe this state or condition. The two concepts refer to the context of lack or deprivation of basic energy services in households. This deprivation is caused by numerous challenges around either accessibility or affordability of basic energy services.

According to Akbari et al. (2001), the concept of energy poverty refers to either a lack of access to electricity or complete dependence on inefficient and polluting resources for energy consumption needs. The latest report from the International Energy Agency (IEA) estimates the number of people without access to electricity worldwide to be around 775 million in 2022 (IEA 2022), which represents a significant rise of about 20 million since 20 years ago. In addition, this report argues that the surge in the number of people without access to electricity results from high prevalence observed in Sub-Saharan Africa (SSA). If not resolved, this condition could be exarcebated by the continued growth in population. Therefore, the IEA projects the total number of people without electricity in 2030 to be around 660 million, of which 85% will be in SSA. Baghdadi et al. (2017) indicated that 30% of the population in Africa do not have access to electricity or rely on traditional energy resources for daily energy requirements. The rate of electricity access in SSA is summarized in Fig. 1.8.

Pye et al. (2015) described energy poverty as a condition whereby people are not able to afford adequate heating to their homes. Pelz et al. (2018) proposed a broader conceptual approach whereby energy poverty could be referred to the inability of a household to meet its domestic energy needs. This definition does also include the inability to provide cooling to households during summer time in areas that experience urban heat island effect (Leal Filho et al. 2021; Hoffman et al. 2020). This condition, which seems to be a growing concern across European countries, comes as a result of a variety of circumstances such as low income, rising energy prices and high energy needs due to poor energy efficiency in buildings. According to Pye et al. (2015): about 54 million (approximately 10.8% of the population) people in the European Union (EU) could not afford to provide heating services to their homes in 2012. Lloyd (2022) reported that more than 35 million people in the EU (approximately 8% of the population) were unable to adequately warm their homes in 2022, and thus facing the "eat or heat dilemma". This author further argues that the fundamental cause of this problem is low income and living in building with

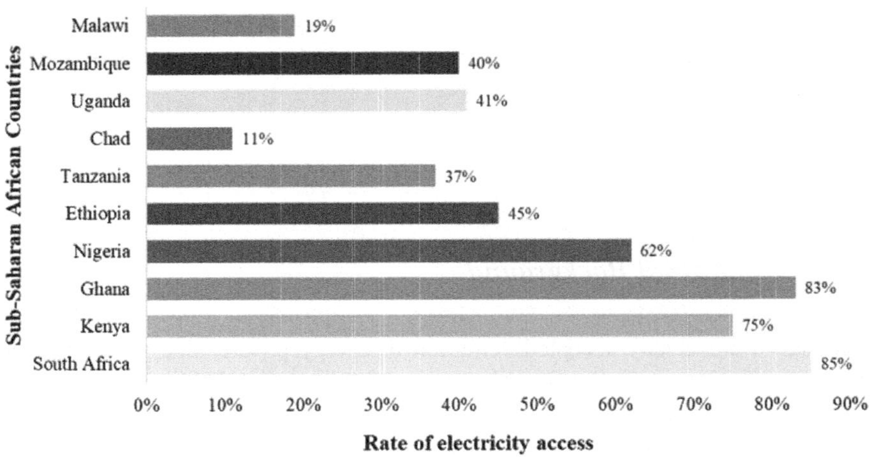

Fig. 1.8 Rate of electricity access in SSA

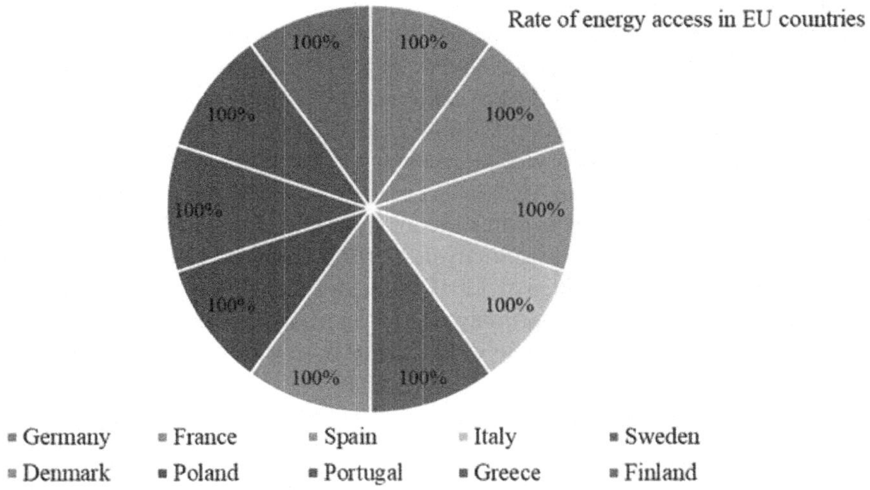

Fig. 1.9 Rate of electricity access in EU Countries

low energy efficiency. Karpinska and Smiech (2020) presented a comprehensive model to assess the scale of exposure to hidden energy poverty in households across 11 Central and Eastern European countries. According to this study, hidden energy poverty is attained if the household income falls below 60% of national median after the housing costs are deducted. The findings reveal that 23.57% of the population of Central and Eastern Europe are exposed to hidden energy poverty. The rate of energy access in EU countries is summarized in Fig. 1.9.

The concept of energy poverty is inherent to socio-economic context of worldwide nations, and could also be regarded as direct consequences of imbalance or misalignment between energy security policies, energy poverty curbing approaches and climate change or low-carbon energy transition governance plans Goldthau (2012).

1.5.2 Socio-Economic Consequences of Energy Poverty

As indicated in IEA (2022), regardless of energy poverty concept, this phenomenon seems to be more prevalent in countries with low gross domestic products (GDP). Therefore, its negative effects could range from socio-economic, environmental, health to education aspects, amongst others:

1. General Well-Being of Communities: communities, where energy poverty is prevalent, experience both social and economic exclusion. This translates into a complete disconnect of these populations from the mainstream society and its economy. Therefore, such communities cannot be participants of general welfare programmes and appear to be marginalised and exposed to starvation (Pondie et al. 2023) or food insecurity (Sola et al. 2016). Social ills (alcohol addiction, human trafficking, child labour, teenage pregnancy, sex work, etc.) tend to become inevitable as a result of survival mode these communities are exposed to.
2. Environmental: the lack of access to clean and affordable energy is ground for continuous reliance on charcoal, paraffin, wood and other high emitting GHG fuels (Bilgili et al. 2022). The use of these fuels contribute to the slow pace of decarbonization and tends to perpetrate deforestation. Therefore, energy poverty is consistent with the degradation of the eco-system, and thus compatible with global warming and climate change.
3. Health: communities under energy poverty may be exposed to detrimental health risks. According to Sefa et al. (2020), increasing energy costs will have an influence on the household budget and may also reflect the performance of the economy. Therefore, these factors could impact on health expenditure and on mental health. Bentley et al. (2023)) reported that the non-affordability of peoples to warm their houses results in significant decline in mental health and show symptoms of depression, anxiety or hypertension. In SSA, Energy deprivation is reported to be the primary source of cumulative stress or anxiety as a result of the discomfort it brings to households with no ability to access energy services (Liddell and Guiney 2015). The IEA (2010) predicted that the number of people using woodstoves across the world is likely to rise from 1.7 billion to 2.8 billion between 2010 and 2030. Furthermore, this institution suggests that death cases, related to this practise, would climb beyond 1.5 million mark by 2030. According to the World Heath Organisation WHO (2018), 3.8 million people are reported to have died from illness related to the use of woodstoves, and among these reported death cases, 27% are due to pneumonia among children under the age of 5; 18% to stroke; 27% to ischaemic heart disease; 20% to chronic obstructive pulmonary disease (COPD); and 8% to lung cancer.

4. Education: according to Sule et al. (2022), energy poverty has an impact on education inequality. Knoth (2015) reported that energy poverty consists of a major stumbling block both at home and at school. He argued that 90 million children in SSA attend primary schools that lack electricity. In Burkina Faso, Cameroon, Malawi and Niger alone, over 80% of primary schools lack access to electricity. This report is concluded by listing 5 key negative impacts of energy poverty on education:

   - significant reduction of teaching resources and classroom materials
   - disturbs work for staff and school administration
   - limited working hours for students to study and complete assignments
   - discourages teachers from working in areas without access to electricity
   - reduces the amount of time that children spend in school.

1.6 Conclusion

This chapter provides an account of the correlation between socio-economic empowerment of humanity and industrialization needs in order to contextualise the interest afforded to fossil-based energy, as well as the subsequent environmental challenges that have been identified (climate change and global warming) as a result of the usage of this form of energy. The set of global measures and imperatives initiated by several UN sustainability conferences, aimed at preserving the ecosystem of the planet, have been extensively discussed in this chapter. This has led to the discussion on the energy transition, which promotes the adoption of renewable energy sources and energy efficient systems for the purposes of achieving decarbonization (zero emissions) or the green economy by 2050. Moreover, this discussion on the critical nature of transition to renewable energy sources has been highlighted as means to build a more sustainable and equitable energy system, and thus, promoting access to clean and affordable energy. The general concepts, such as attributed to energy poverty, have been debated alongside the socio-economic consequences or impacts of this condition. Therefore, while it is generally argued that energy poverty is not compatible with socio-economic expansion or development of humanity, it is equally not conducive to the protection and safeguard of the planet ecosystem.

References

Ahmed, A., Gasparatos, A.: Multi-dimensional energy poverty patterns around industrial crop projects in Ghana: enhancing the energy poverty alleviation potential of rural development strategies. Energy Policy **137**, 111123 (2020). https://doi.org/10.1016/j.enpol.2019.111123

Akbari, H., Pomerantz, M., Taha, H.: Cool surfaces and shade trees to reduce energy use and improve air quality in urban areas. Sol. Energy **70**(3), 295–310 (2001)

Baghdadi, I., El Yaakoubi, A., Attari, K., Leemrani, Z., Asselman, A.: Performance investigation of a PV system connected to a grid. In: 11th International Conference Interdisciplinarity in Engineering, 5-6 October, Tirgu-Mures, Romania (2017)

Bentley, R., Daniel, L., Li, Y., Baker, E., Li, A.: The effect of energy poverty on mental health, cardiovascular disease and respiratory health: a longitudinal analysis. The Lancet Regional Health - Western Pacific (2023)

Bilgili, F., Ozturk, I., Kocak, E., Kuskaya, S., Cingoz, A.: The nexus between access to electricity and CO_2 damage in Asian countries: the evidence from quantile regression models. Energy and Buildings **256** (2022). https://doi.org/10.1016/j.enbuild.2021.111761. https://www.sciencedirect.com/science/article/pii/S0378778821010458

Bressler, S.L.: A historical overview of the development of programmable logic controllers. IEEE Trans. Educ. **44**(2), 170–174 (2001)

Brynjolfsson, E., McAfee, A.: Race Against the Machine: How the Digital Revolution is Accelerating Innovation, Driving Productivity, and Irreversibly Transforming Employment and the Economy. Digital Frontier Press (2011)

Castaño-Rosa, R., Solís-Guzmán, J., Marrero, M.: Energy poverty goes south? Understanding the costs of energy poverty with the index of vulnerable homes in Spain. Energy Res. Soc. Sci. **60**, e101325 (2020). https://doi.org/10.1016/j.erss.2019.101325

Chen, L., Msigwa, G., Yang, M., Osman, A.I., Fawzy, S., Rooney, D.W., Yap, P.: Strategies to achieve a carbon neutral society: a review. Environ. Chem. Lett. **20**, 2277–2310 (2022). https://doi.org/10.1007/s10311-022-01435-8

Chiu, Y.H., Chu, C.W.: Life cycle greenhouse gas emissions and the environmental impacts of coal and natural gas power plants. Appl. Energy **285**, 116533 (2021)

Cleveland, C.J.: Energy and Civilization: A History. Cambridge University Press (2019)

Crafts, N.F.: British Economic Growth During the Industrial Revolution. Oxford University Press (1985)

de La Paz, P.T., Tárrega, F.J., Su, Z., Monllor, P.: Sources of energy poverty: a factor analysis approach for Spain. Front. Energy Res. **10**, 847845 (2022). https://doi.org/10.3389/fenrg.2022.847845

Emerald, V.C.: Three pillars of the energy transition: green electrons, molecules and bits (2023). https://emerald.vc/three-pillars-of-the-energy-transition-green-electrons-molecules-and-bits

Fuller, R., Elliott, R., Cowie, H., Schiek, W. (Eds.).: Environmental Pollution: A Comprehensive Introduction. Routledge (2020)

Gately, D.: The global energy mix: Fossil fuels, nuclear and renewable sources. J. Econ. Perspect. **29**(1), 135–156 (2015)

Goldthau, A.: From the state to the market and back: policy implications of changing energy paradigms. Glob. Policy **3**, 198–210. https://doi.org/10.1111/j.1758-5899.2011.00145.x (2012)

Goldthau, A., Sovacool, B., Benjamin, K.: the uniqueness of the energy security, justice, and governance problem. Energy Policy **41**, 232–240 (2011). https://doi.org/10.1016/j.enpol.2011.10.042

Grinin, L.E., Korotayev, A.V., de Munck, V. (Eds.).: Energy and Human Resource: Case Studies from Europe, Africa, Asia, and America. Cambridge Scholars Publishing (2015)

Hansen, J., Sato, M., Ruedy, R.: Perception of climate change. Proc. Natl. Acad. Sci. **109**(37), E2415–E2423 (2012)

Held, I.M., Soden, B.J.: Robust responses of the hydrological cycle to global warming. J. Clim. **33**(12), 4951–4962 (2020)

Herring, H.: The industrial revolution and economic growth. Oxf. Rev. Econ. Policy **26**(4), 561–585 (2010)

Hoffman, J.S., Shandas, V., Pendleton, N.: The effects of historical housing policies on resident exposure to intra-urban heat: a study of 108 US urban areas. Climate. **8**(1), 12 (2020). https://doi.org/10.3390/cli8010012

Hsiang, S.M., Burke, M., Miguel, E.: Quantifying the influence of climate on human conflict. Science **341**(6151), 1235367 (2017)

IEA: Energy poverty: how to make modern energy access universal? (2010). https://www.undp.org/content/undp/en/home/librarypage/environment-energy/sustainable-energy/energypoverty-how-to-make-modern-energy-access-universal.html

International Energy Agency (IEA): SDG7: Data and Projections (2022). https://www.iea.org/reports/sdg7-data-and-projections

International Renewable Energy Agency (IRENA): World Energy Transitions Outlook (2022). https://www.irena.org/Digital-Report/World-Energy-Transitions-Outlook-2022

Jackson, R.B., Canadell, J.G., Le Quéré, C., Andrew, R.M., Korsbakken, J.I., Liu, Z. ...and Peters, G.P. : Reaching peak emissions. Nat. Climate Change **10**(3), 196–199 (2020)

Jones, M.W., Peters, G.P., Gasser, T., et al.: National contributions to climate change due to historical emissions of carbon dioxide, methane, and nitrous oxide since 1850. Sci Data **10**, 155 (2023)

Kallis, G., Kerschner, C., Martinez-Alier, J.: The economics of degrowth. Ecol. Econ. **84**, 172–180 (2012)

Karpinska, L., Śmiech, S.: On the persistence of energy poverty in Europe: how hard is it for the poor to escape? Energy Res. Lett. **1**(3) (2020). https://doi.org/10.46557/001c.14158

Knoth, G.: A classroom's worst nightmare? Energy poverty (2015). https://www.one.org/us/blog/a-classrooms-worst-nightmare-energy-poverty

Landes, D.S.: The Unbound Prometheus: Technological Change and Industrial Development in Western Europe from 1750 to the present. Cambridge University Press (2003)

Leal Filho, W., Wolf, F., Castro-Díaz, R., Li, C., Ojeh, V.N., Gutiérrez, N., Nagy, G.J., Savić, S., Natenzon, C.E., Quasem Al-Amin, A., et al.: Addressing the urban heat Islands effect: a cross-country assessment of the role of green infrastructure. Sustainability **13**, 753 (2021). https://doi.org/10.3390/su13020753

Liddell, C., Guiney, C.: Living in a cold and damp home: Frameworks for understanding impacts on mental well-being. Public Health **129**, 191–199 (2015). https://doi.org/10.1016/j.puhe.2014.11.007

Lloyd, N.: Turning up the heat on Europe's fuel poverty crisis (2022). https://www.euronews.com/next/2022/01/26/turning-up-the-heat-on-europe-s-energy-poverty-crisis

Lu, Y., Khan, Z.A., Alvarez-Alvarado, M.S., Zhang, Y., Huang, Z., Imran, M.A.: Critical review of sustainable energy policies for the promotion of renewable energy sources. Sustainability **12**, 5078 (2020)

Manisalidis, I., Stavropoulou, E., Stavropoulos, A., Bezirtzoglou, E.: Environmental and health impacts of air pollution: a review. Front. Public Health **8**, 14 (2020)

Mathur, R.B., Rao, N. S.: Handbook of Clean Energy Systems, Vol. 6: Fossil Energy Technologies. Wiley (2018)

Mokyr, J.: Why "More Work for Mother?": knowledge and household behavior, 1870–1945. J. Econ. Hist. **59**(1), 1–41 (1999)

Morris, I.: Why the West Rules–for Now: The Patterns of History, and What They Reveal about the Future. Macmillan (2010)

Mountford, H., Waskow, D., Gonzalez, L., Gajjar, C., Cogswell, N., Holt, M., Fransen, T., Bergen, M., Gerholdt, R.: COP26: Key Outcomes From the UN Climate Talks in Glasgow (2021). https://www.wri.org/insights/cop26-key-outcomes-un-climate-talks-glasgow

National Oceanic and Atmospheric Administration (NOAA): Carbon Dioxide Now More than 50% Higher than Pre-Industrial Levels. Available online: https://www.noaa.gov/news-release/carbon-dioxide-now-more-than-50-higher-than-pre-industrial-levels (2022)

Nilkar, A.S., Orme, C.J., Klaehn, J.R., Zhao, H., Adhikari, B.: Life cycle assessment of innovative carbon dioxide selective membranes from low carbon emission sources: a comparative study. Membranes, MDPI **13**, 410 (2023)

OECD/IEA Climate Change Expert Group: Aligning short-term climate action with long-term climate goals: Opportunities and options for enhancing alignment between NDCs and long-term strategies, OECD Publishing, Paris (2020). http://www.oecd.org/environment/cc/ccxg.htm

Pelz, S., Pachauri, S., Groh, S.: A critical review of modern approaches for multidimensional energy poverty measurement. Wiley Interdiscip. Rev. (Energy and Environment) **7**(6) (2018). https://doi.org/10.1002/wene.304

Pondie, T.M., Engwali, F. D., Nkoa, B.E.O., Domguia, E.N.: Does energy poverty increases starvation? Evidence from sub-Saharan Africa. Environ. Sci. Pollut. Res. Int. **30**(17), 8721–48738 (2023). https://doi.org/10.1007/s11356-023-25997-4. Epub 2023 Mar 4. PMID: 36869173; PMCID: PMC9984292

Pye, S., Dobbins, A., Baffert, C., Brajkovic, J., Deane, P., De Miglio, R.: Addressing Energy Poverty and Vulnerable Consumers in the Energy Sector Across the EU: analysis of policies and measures. L' Europe en Formation **4**(378), 64–89 (2015). https://www.cairn.info/revue-l-europe-en-formation-2015-4-page-64.htm

Ripple, W.J., Wolf, C., Newsome, T.M., Barnard, P., Moomaw, W.R.: World scientists' warning of a climate emergency. Bioscience **71**(3), 227–232 (2021)

Ritchie, H., Roser, M., Rosado, P.: Energy. Our World in Data (2022). https://ourworldindata.org/energy

Sachs, J.D., Schmidt-Traub, G., Kroll, C., Lafortune, G. Fuller, G.: Sustainable Development Report 2019: Transformations to Achieve the Sustainable Development Goals. Bertelsmann Stiftung and Sustainable Development Solutions Network (2019)

Santos, F.D., Ferreira, P.L., Pedersen, J.S.T: The climate change challenge: a review of the barriers and solutions to deliver a paris solution. Climate **10**(05) (2022). https://doi.org/10.3390/cli10050075

Schiffer, H.P., Heinemann, T., Franzinelli, M. (Eds.).: Power Generation from Solid Fuels. Elsevier (2018)

Sefa, C.A., Smyth, R., Farrell, L.: Fuel poverty and subjective wellbeing. Energy Econ. Elsevier **86**(C) (2020). https://doi.org/10.1016/j.eneco.2019.104650

Smail, E.: Energy Poverty: Causes, Effects and Solutions (2023). https://energytracker.asia/energy-poverty-causes-effects-solutions

Smil, V. : Energy and Civilization: A History. MIT Press (2017)

Sola, P., Ochieng, C., Yila, J., et al.: Links between energy access and food security in sub Saharan Africa: an exploratory review. Food Sec. **8**, 635–642 (2016). https://doi.org/10.1007/s12571-016-0570-1

Sorensen, B.A.: History of Energy: Northern Europe from the Stone Age to the Present Day. Routledge, London, UK (2013). ISBN 1136487891

Sovacool, B.K.: Contesting the Future of Nuclear Power: A Critical Global Assessment of Atomic Energy. World Scientific (2011)

Sule, I.K., Yusuf, A.M., Salihu, M.K.: Impact of energy poverty on education inequality and infant mortality in some selected African countries. Energy Nexus **5**, 100034 (2022). https://doi.org/10.1016/j.nexus.2021.100034

UN Environment Programme (UNEP): Towards a Green Economy: Pathways to Sustainable Development and Poverty Eradication. Available at: https://www.unep.org/resources/report/towards-green-economy-pathways-sustainable-development-and-poverty-eradication (2011)

UN Framework Convention on Climate Change (UNFCCC): The Paris Agreement. https://unfccc.int/sites/default/files/resource/parisagreement-publication (2016)

UN Sustainable Development Summit (SDS): Sustainable development and disaster risk reduction (2015). http://www.un.org/sustainabledevelopment/summit

UNDP: Climate Promise (2022). https://climatepromise.undp.org/news-and-stories/what-just-transition-and-why-it-important

United Nations (UN): The 2030 Agenda and the Sustainable Development Goals: An opportunity for Latin America and the Caribbean, Santiago (2018). https://repositorio.cepal.org/bitstream/handle/11362/40156/S1801140

United Nations Department of Economic and Social Affairs (DESA): Transforming our world: the 2030 Agenda for Sustainable Development (2015). http://www.sdgs.un.org/2030agenda

United Nations Development Programme (UNDP): Greening Development: A Guide for Country-Level Implementation. https://www.undp.org/publications/greening-development-guide-country-level-implementation (2020)
United Nations Development Programme (UNDP): Human Development Report 2020: The Next Frontier - Human Development and the Anthropocene. UNDP Publications (2020)
United Nations Development Programme (UNDP): Tracking SDG 7: The Energy Progress Report (2020). https://trackingsdg7.esmap.org
United Nations Industrial Development Organization (UNIDO): Industrial Energy Transition: Pathways towards Sustainable Energy for All. UNIDO Publications (2021)
United States Environmental Protection Agency (EPA): Sources of Greenhouse Gas Emissions (2021). Retrieved from http://www.epa.gov/ghgemissions/sources-greenhouse-gas-emissions
Wang, X., Lo, K.: Just transition, energy transition, low-carbon transition, conceptual review. Energy Res. Soc. Sci. **82**, 102291 (2021). https://doi.org/10.1016/j.erss.2021.102291
Wang, J., Gao, Y., Huang, Y., Liu, Z., Zhang, Y.: A review of fossil fuel-fired power plant emission standards. J. Clean. Prod. **244**, 118575 (2020)
WHO: Household air pollution and health (2018). https://www.who.int/news-room/fact-sheets/detail/household-air-pollution-and-health

Chapter 2
Assessment Methods of Energy Poverty

2.1 Introduction

The existing body of literature on energy poverty strongly indicates that at its core, this condition involves the deprivation of essential energy services for households. The extent of the incidence of this type of poverty on households and its occupants justifies the attention currently afforded to this problem. Therefore, energy poverty is highlighted as a priority target based on goals 1 and 7 of the UN 17 SDGs (SDS 2015), as well as on the creation of the EU Energy Poverty Observatory (EPOV) (EU 2020). However, this form of poverty is difficult to measure, and therefore, requires careful consideration of the various determinants (low income, energy inefficient dwellings or buildings, high cost of energy, households' needs, etc.), which are associated with energy poverty vulnerability in the context of affordability such as experienced in advanced economies (Bouzarovski and Tirado Heredo 2017; Bouzarovski and Petrova 2015; Thomson et al. 2017). Additional determinants as well as respective variants (socio-economic, demographic, etc.) of energy deprivation, which could aggravate the vulnerability of households, should be considered given the multidimension nature of this condition in SSA countries: both affordability and accessibility (Bouzarovski and Petrova 2015; Boqiang and Okyere 2020). Hence, access to electricity for households, individual factors (gender, race, educational attainment, etc.) and household characteristics such as: electrified, geographical location, size, number of occupants, etc. should be of great value (Ismail and Khembo 2015). Therefore, the validity of proposed assessment or evaluation methods of energy poverty should be based on the extent of the causal link measured between energy deprivation and related determinants as well as their associated factors. In this light, a review of the various energy deprivation assessment techniques, proposed in the literature, is conducted in this section of the book. Also, an in-depth analysis of available techniques, for curbing energy poverty, will be attempted prior to concluding this chapter.

P. N. Bokoro and G. Sharma, *Curbing Energy Poverty in Sub-Saharan Africa*,
https://doi.org/10.1007/978-3-031-90072-3_2

2.2 Review of Energy Poverty Assessment Methods

Assessment methods of energy poverty are strictly related to the conceptual understanding of this condition. Therefore, estimation metrics proposed in the literature focus either on affordability, on accessibility or on both.

2.2.1 Affordability Metrics

According to Thomson (2013), three fundamental approaches are generally used for the measurement of the affordability dimension of energy poverty. These approaches are termed as temperature, expenditure and consensual. In support of Thomson's work, Barrella et al. (2022) formulated a classification of approaches based on the following three groups: direct approach, which encompasses temperature or thermal comfort to model energy poverty, the income-expenditure (otherwise referred to as objective) approach, which is based on energy consumption and household's revenues, and the self-reported (otherwise referred to as subjective) approach. However, Rademaekers et al. (2016a) reported on the fourth approach of evaluating energy poverty. This was later supported by Thema and Vondung (2021). In this section, a critical analysis of these approaches is undertaken.

1. Direct Approach:
   This assessment approach is based on measured internal room temperature data, which are used to determine whether or not households are capable of maintaining the threshold levels of warmth. While this assessment model may appear to be quite simple, it quickly triggers the debate on the generally accepted or adopted threshold temperature in households. Therefore, lower temperature maintained in some households may not be related to financial constraints.
   Boardman (2010); Healy (2004) and Critchley et al. (2007) provided a detailed account of the relative nature of thermal comfort and financial constraints in some parts of the United Kingdom (UK). Therefore, Healy and Clinch (2002) made use of random probability-based sampling to conduct temperature measurement on 1500 households, and found that 29.4% of energy poor households were able to maintain a living-room temperature of 18 °C or less, as opposed to 8.8% of other households. This finding suggests that there is a mismatch between energy poor households and those experiencing thermal discomfort. Therefore, the main disadvantage of this model lies in the fact that indoor thermal comfort does not constitute a good indicator of energy poverty. This has made this method less popular and not reliable enough to be used.
2. Income-Expenditure or Objective Approach:
   This assessment approach relies on the ratio between household income and energy expenditure. Therefore, a household is classified as energy poor if its energy expenditure exceeds its revenue by a threshold amount (Waddams et al. 2007; Fahmy 2011). However, this assessment metric is confronted with two

important considerations: the estimation of energy expenditure as well as the determination of household income thresholds.

In an attempt to provide guidelines over the correct selection of expenditure thresholds, Boardman (2012) proposed the concept of absolute and relative expenditure thresholds. According to this author, absolute expenditure threshold whereby disproportionate energy expenditure is measured on the basis of a minimum income standard (MIS) (Antepara et al. 2020; Papada and Kaliampakos 2020; Faiella and Lavecchia 2020), and relative expenditure threshold whereby the median cost to income ratio for all households is applied (Meyer et al. 2018; Betto et al. 2020; Karpinska and Smiech 2020).

For the determination of adequate household income thresholds, Boardman proposed the following measures: before housing costs or after housing costs, disposable income, and equivalised or adjusted income, which include households sizes and composition. According to Fahmy (2011), failure to apply equivalisation of households' incomes could lead to the over-estimation of available incomes to larger households to meet their domestic energy requirements. As generally accepted measure, the 10% indicator suggests that a household is classified to be energy-deprived if it consumes over 10% of its total income on energy.

Several studies conducted in the UK as well as in other countries in Europe supported this indicator (Papada and Kaliampakos 2016; Legendre and Ricci 2015; Scarpellini et al. 2015; O'Sullivan et al. 2011). Mould and Baker (2017) argued that about 30% of households in Scotland were classified as energy-poor when this measure was applied in 2015. Bennett et al. (2022) applied this measure and found the correlation between energy payment methods, income, type and composition of the household. Bradshaw (2008) reported that an estimate of 7% of households spent more than 10% of their net incomes on energy between 2005 and 2006.

Hills (2012) argued that a household is classified as energy poor if its income after energy services expenditure falls below the poverty threshold and energy expenditure is higher than the median cost of the population. Costa-Rosa et al. (2019) proposed the 2 M indicator, which evaluates a houseold as energy poor if a household spends more than double the median, double the mean, double the median share, and double the mean share of its income on energy services. Hills (2011) introduced the After Fuel-Cost Poverty (AFCP) indicator, which defines a household as energy-poor, if it does not have enough income to spend on essential energy services, after covering housing and other basic needs. However, the validity of this approach remains questionable since low absolute energy expenditure may not be attributed to under-consumption, and median-based indicators could be misleading. Furthermore, Income-Expenditure metrics do not take into account households patterns or characteristics.

3. Self-Reported or Consensual (Subjective) Approach:

   Owing to the drawbacks associated to income-expenditure approach, self-reported or consensual methods were proposed to model energy poverty. This method focusses on consensual approach such as advocated by Gordon et al. (2000). This approach takes into account several dimensions or subjective indicators of energy poverty (Thema and Vondung 2021), such as:

- Ability to keep homes adequately warm/cool
- Arrears on energy or utility bills
- Building conditions (dampness, mould, etc.)

By applying the ability to adequately warm households, Healy and Clinch (2002) reported that 12.7% of Irish households experienced occasional energy poverty, whilst a further 4.7% suffered persistent energy poverty, which represent approximately 227,000 households in total. In addition, the Poverty and Social Exclusion (PSE) project, which consisted of a mix of various aspects related to energy poverty (Gordon et al. 2013), could be implemented in the following three stages: attitudes survey to determine necessities; a living standards survey in order to identify households who lacked necessities by choice and those who could not afford these necessities; and lastly a deprivation count to determine how many people cannot afford groups of necessities. This project concludes that 96% of adults believed that heating to warm living areas of a household is a basic requirement, and 94% considered a damp-free household a basic requirement. Therefore, the project showed that 9% of UK households (2.3 million) cannot afford to heat living areas of their home and 10% of households (2.7 million) live in a damp home. The most important benefit of this approach resides in its ability to integrate the wider indicators of energy poverty, such as social exclusion and material deprivation. However, the major criticism in the use of subjective indicators remains the error of exclusion, whereby households may or may not be identified as energy poor under different measures or indicators (Dubois 2012).

4. Outcome-based Approach:
   This assessment approach quantifies energy poverty by attempting to measure if actual energy consumption is sufficient to attain an adequate standard of living (Halkos and Kostakis 2023). It is fundamentally anchored on a group of indicators, which provides a proxy for energy poverty based on outcomes. This model focusses on the use of utility data such as arrears, or health outcomes, excess mortality in extreme climatic conditions, etc. However, the complexity to correctly identify and measure energy poverty outcomes make this assessment model is quite difficult.

The approaches discussed above are much more convenient to measure the affordability dimension of energy poverty, which is reported to be more prevalent in developed countries. However, as suggested by Cong et al. (2022), income-expenditure approaches may fall short of classifying hidden poverty condition, whereby households may be spending small amounts of their income on energy bills due to deliberate supply interruptions aimed at managing consumption.

2.2.2 Accessibility Metrics

According to Culver (2017), three fundamental approaches (based on single indicator) are generally used to measure the accessibility dimension of energy poverty. These metrics are given as follows: energy access, energy inputs and the outcomes of energy use.

1. Energy Access:
   This assessment approach is based on the measurement of households' access to energy services (electricity, cooking fuels, etc.). Therefore, the percentage amount of households with connection to an electric grid (electrification rate), and access to traditional fuels (fuelwood, charcoal, crop residues, etc.) consist of major indicators. Furthermore, the IEA (2020) emphasizes the link between energy access and its crucial role in enabling productive economic activities and providing essential public services. The binary nature of this metric makes it quite convenient for energy poverty measurement and data collection. However, this metric is confronted to the following drawbacks:

   - Energy access metric is restricted to household energy usage
   - Energy access metric is not inclusive of energy quality
   - Inconsistent alignment between energy and income poverty in rural and urban areas

2. Energy Inputs:
   Owing to the shortcomings related to the use of energy access metric, the measurement of energy inputs has been relied upon. This approach assesses energy poverty in terms of energy consumed in a household using units of energy. This method is based on the assumption that providing energy input to a household automatically ensures the availability of energy services. Hence, to accurately measure the energy consumed in a household, it is essential to consider the actual utilisation of primary, final and useful energy.
   According to Goldemberg and Johansson (1995), energy poverty in the context of primary energy utilisation can be determined by estimating the average per capita national energy consumption using the Human Development Index (HDI). Foster et al. (2000) argued that the average final energy consumption of households could be used to measure the energy poverty line. Goldemberg et al. (1987) estimated the minimum energy requirements using a back calculation of the energy that would be needed to supply a normative set of basic energy services. Pachauri and Spreng (2004) introduced the Energy Access-Consumption Matrix, which systematically integrates the physical availability of various energy types with their corresponding consumption levels. This enables the distribution of useful energy to be established.
   While this approach could be of great assistance in the process of formulating and comparing national and global policies for energy demand, however, the concept of poverty line such as evoked in this approach may change in time and geography as a result of improvements in the efficiency of appliances and the energy.

3. Outcomes of Energy Use:
   This approach is based on the fact that unfavourable outcomes will be experienced should households have no access to energy services. Therefore, the causality between poor or bad services experienced and energy poverty consists of the major focus of this method. In this light, three fundamental metrics have been proposed to measure the outcomes of energy poverty (Mirza and Szirmai 2010; Sen 1999):

   - Individual effects of energy use that are indicators of energy poverty (health, environmental, etc.)
   - Opportunity costs (time lost, etc.)
   - Available choice

   However, the measurement of these metrics could be problematic and may lead to difficulties in collecting data.

2.2.3 Composite Metrics

According to Drago and Gatto (2023), composite metrics of energy poverty consist of aggregated indices or indicators, which are able to produce a unique quantification of this condition. This measure could be used to provide a comparison of different statistical units, deemed useful for assessment or evaluation of energy access. These metrics attempt to assess the several dimensions of energy poverty in a single number. Siksnelyte-Butkiene et al. (2021) provided an account of most commonly used composite metrics: the Multi-Tier Framework (MTF), the Household Energy Access Index (HEAI), the Comprehensive Energy Poverty Index (CEPI), the Energy Development Index (EDI), and the Multi-dimensional Energy Poverty Index (MEPI).

1. Multi-Tier Framework (Quality of energy delivered):
   This approach was introduced by the World Bank's Energy Sector Management Assistance Programme (ESMAP) in 2013. It does assess the attributes of energy supply in order to measure the energy services required in households and business entities (Bhatia and Angelou 2015). It combines several indicators in order to compute an aggregation using averaging. Therefore, the MTF assesses the quality of energy in the following domains: household energy, energy for productive uses, and community energy. The MTF is the most advanced measure of energy poverty as it comes closest to approximating energy services. This metric assumes that energy services impose the need for quality of energy, which could be described by the following attributes: capacity, duration, quality, reliability, affordability, legality, and health and safety.
   The application of the MTF in the household energy (electricity) domain for the above-described attributes is indicated in Table 2.1.
   According to Table 2.1, Tier 1 access to electricity services requires the provision of at least 1,000 lumen hours and cell phone charging for at least four hours a day,

Table 2.1 Five tiers of access to electricity services (Bhatia and Angelou 2015; Culver 2017; Mullen and Wade 2020)

Attributes	Tier 0	Tier 1	Tier 2	Tier 3	Tier 4	Tier 5
Peak capacity (W)	–	>3	>50	>200	>800	>2000
Resulting (Wh) per user	–	>12	>200	>1000	>34000	>82000
Services	–	1000 lumens hr/day	lighting, aircon, TV, charging	–	–	–
Duration (hr)	–	≥ 4	≥ 4	≥ 8	≥ 16	≥ 23
Evening supply (hr)	–	≥ 1	≥ 2	≥ 3	≥ 4	≥ 4
Reliability (number of interruptions per week)	–	–	–	–	14	3
Quality	–	–	–	–	Yes	Yes
Affordability	–	–	–	Yes	Yes	Yes
Legality	–	–	–	–	Yes	Yes
Health and safety	–	–	–	–	Yes	Yes

of which at least one hour must be in the evening. On the other end, Tier 5 consists of 23 h of reliable electricity provision to households, at a legal cost of less than 5% of household income. In addition to the afore-mentioned attributes associated with electricity quality, the multi-tier matrix for electricity consumption (either on annual or daily basis) as well as for access to energy-intensive services (productive services) could be considered. In this case, the overall tier assignment is estimated using the lowest tier assignment of each of the attributes among the assessments. Such assessments may be aggregated in a simple mathematical average as an index associated with access in each domain, which are further aggregated to determine the overall energy access index.

The significant advantage of this assessment metrix is such that it does include all aspects of the UN SDGs, and thus could be applied as technical grounds for identifying energy access gaps, tracking or monitoring progress towards the attainment of universal access to affordable, reliable and modern energy services, and for the purposes of developing policy directions and providing guidance for investment decisions. However, Groh et al. (2016) made use of a case study of 230 households in Bangladesh to develop a set of arguments against the MTF for electricity access. These authors argue that while both electricity supply and household electricity poverty consist of multi-faceted concepts and, therefore, require separate metrics. Furthermore, the understanding of individual dimensions and their aggregation method may also require alternative approaches. In addition, Siksnelyte-Butkiene et al. (2021) observed that the MTF partially consists of a non-compensatory composite indicator.

2. Household Energy Access Index (HEAI):
   This multi-dimensional framework for measuring energy access, mainly in developing countries, was proposed by Tait (2017). This technique consolidates the attributes or indicators, previously used in the ESMAP approach, into the following four indicators: fuel use (variables that indicate type of electricity accessed and fuels consumed), affordability (expenditure as a proportion of income), safety (identified risk factors), and reliability (duration of interruptions experienced) in order to estimate the household energy access index. Furthermore, in this approach, the indicators are associated with the level of the household and not to energy services. The development and contextual application (South African context is used) of the four indicators of this composite metric are conducted as follows:

   - Fuel Use:
     In the South African context, the following hierarchical access is to be assigned to a given household: no access to electricity (0), informal or illegal electricity connection (1) and metered electricity connection and usage of inadequate fuels (3). Therefore, the overall score for a given location of n households is determined as follows:
     $$Score = \frac{\sum(P \times S)}{3} \tag{2.1}$$
     where P is the proportion of households and S is the category scale.
   - Affordability:
     This indicator is closely related to variables such as household's income and the costs of energy services. Although, these variables (income and costs of energy services) may consist of objective variables, there is however some measure of subjectivity, which is usually associated with reporting criteria (ability to keep homes warm/cool; arrears on energy bills, etc.). Therefore, based on the South African context such as developed by Tait (2017), whereby the income/expenditure (objective) and self-reported (consensual or subjective) methods were applied to assess the affordability of household's energy needs in two areas of Cape Town, South Africa. The scores for the affordability indicator are given in Table 2.2.

Table 2.2 Assessment of scores for each applied method

Objective method		Subjective method	
1	< 10%	1	Yes
2	10–20%	2	Sometimes
3	20–30%	3	No
4	>30%		

For the objective method, the score is determined using the following equation:

$$Score = \frac{\sum(P \times S^2)}{4}, \tag{2.2}$$

Similarly for the subjective method, the score is determined using the following equation:

$$Score = \frac{\sum(P \times S^2)}{3} \tag{2.3}$$

where P is the proportion of households and S is the category scale.

- Safety Indicator:
  This indicator estimates the potential safety risks or energy-related hazards (environmental pollution, burns, fires, electric shocks, poisonings, etc.) related to the use of energy sources in households. In the South African context, burns and fires are more prevalent energy risks. However, a number of aggravating factors to energy risks should be noted: housing materials, inadequately small homes, unplanned settlement layouts, etc. Therefore, five categories of risk factors are used in this context: use of unsafe fuels, indoor air pollution, lack of fire safety knowledge, electrical safety risks and behavioural risk. The safety risk score could be assessed using the following equation:

$$Score = \sum(P \times R_n) \tag{2.4}$$

  where P is the proportion of households and R_n is the number of risk factors.
- Reliability Indicator:
  Generally, reliability is associated with the number and duration of service interruptions experienced by households or consumers. However in this case, reliability indicator does measure economic, financial and/or social costs of the impacts of service interruptions or outages on households.
  Seuret-Jimenez et al. (2020) applied Fuzzy logic in order to achieve a three variable (average expenditure on monthly transport, electricity, and cooking fuel) index. The proposed index consists of similar indicators with the metric proposed by Tait (2017).

3. Comprehensive Energy Poverty Index (CEPI):
   This approach was proposed by Khanna et al. (2019) in a bid to determine the magnitude of energy poverty and to identify micro-level solutions in South and Southeast Asia. In terms of this metric, energy poverty is computed on the basis of the following: availability, accessibility, and affordability. Therefore, the three-dimensional concept of energy poverty is introduced through this composite metric. This technique is more applicable in developing countries, and is based on the following four indicators: access to electricity, clean fuels, technologies for cooking, total energy supply, and total final energy consumption. Therefore, to

compute the extent of energy, the weighted average of the four energy poverty indicators (WAEPI) per country per year is introduced as follows:

$$WAEPI_{c,year} = \sum \left(w_1 \cdot X_{e_n} + w_2 \cdot X_{mf_n} + w_3 \cdot k^1_{pc_n} + w_4 \cdot k^2_{pc_n}\right) \tag{2.5}$$

where: w_1, w_2, w_3 and w_4 are the assigned weights to the four indicators, X_{e_n} represents normalised access to electricity, X_{mf_n} indicates the normalised access to modern fuel, $k^1_{pc_n}$ is the total final energy consumption per capita, and $k^2_{pc_n}$ is the total primary energy supply per capita. Therefore, the CEPI metric per country is estimated as follows:

$$CEPI = 100 - WAEPI \tag{2.6}$$

Therefore, the CEPI metric is in fact similar to the model introduced earlier by Ayodele et al. (2018), which estimates the composite electrical energy poverty index (CEEPI). The CEEPI metric resulted from an evaluation of energy poverty among micro-enterprises in Ibadan (Nigeria) using the three following indices: suitability, electricity affordability, and electricity availability index. In this approach, the indices are computed in a scale of 0–1, with 0 being the least electrical energy poverty (EEP) and 1 being the acute EEP.

4. Energy Development Index (EDI):
This metric was developed by the IEA and does focus on a country's or region's transition to the use of modern fuels specifically for developing countries. Malla (2013) applied this composite metric to study energy access and poverty across thirteen regions of Nepal. It consists of an arithmetic mean of the following four indicators: per capita commercial energy consumption, per capita consumption of electricity in the residential sector, share of modern fuels in total residential energy sector consumption, and share of population with access to electricity. Each of the above indicators could be determined as an index between 0 and 1. Therefore, the following relationship is applied:

$$I_i = \frac{V_a - V_{min}}{V_{max} - V_{min}} \tag{2.7}$$

where I_i is the ith indicator, V_a is the actual country value, V_{min} is the observed minimum value of the indicator for the country covered, and V_{max} is the corresponding maximum value.
Therefore, the EDI composite metric could be estimated as follows:

$$EDI = \frac{\sum I_i}{4} \tag{2.8}$$

5. Multi-dimensional Energy Poverty Index (MEPI):
This is one of the most popular composite metric, which was introduced by Sokołowski et al. (2020). It mixes up both objective (low income and high actual

costs) and subjective (inability to keep home adequately warm, housing faults and bill difficulties) indicators to compute energy poverty specifically in developed countries. It does suggest that a household suffers from energy poverty if at least two of the five indicators attest to that fact. Nussbaumer et al. (2012, 2013) suggested that this composite metric provides not only the incidence and intensity of energy poverty, but consists of robust and adaptable approach suitable to quantify national contexts and policy-based objectives. In this context, a proposed description of the dimensions and indicators used in the MEPI is summarised in Table 2.3.

According to Rizal et al. (2024), the variables characterising the indicators of the MEPI's five dimensions of energy poverty (see Table 2.3) could be expressed either as monetary or non-monetary. Hence, the concept of monetary and non-monetary objective indicators associated with these authors' approach to the MEPI. Therefore, non-monetary objective variable types are assigned to availability, affordability, energy consumption and energy services in accordance with previously reported studies (Kaygusuz 2011), Rademaekers et al., 2014, (Hartono et al. 2020; Qurat-ul-Ann and Mirza 2020). On the other end, monetary objective variable types are assigned to the accessibility dimension (Nussbaumer et al.

Table 2.3 Dimensions and Indicators of the MEPI (Rizal et al. 2024)

Dimension	Indicators	Description
1. Availability	% of households with or without access to electricity and/or clean cooking fuels	Modern energy facilities in the same geographic location
2. Accessibility	Electricity access, LPG access	Main light source, main cooking fuel
3. Affordability	TPR, 2 M, M/2, LIHC, LI-L/HC	Energy expenditure, a portion of energy expenditure, a portion of energy expenditure, disposable income, relative or absolute energy, expenditure, disposable income
4. Energy Consumption	Per capita electricity consumption, per capita cooking fuel consumption	Electricity consumption, cooking fuel consumption
5. Energy services through household appliances, for:		
a. Processing food and beverages (Primary)	≥ 5.5 kg gas cylinders ownership, Refrigerator ownership	Household appliance ownership, Household appliance ownership
b. Education, Entertainment, Information & Communications (secondary)	≥30-inch flat television ownership, computer ownership (PC/Laptop), HP ownership	Household appliance ownership, Household appliance ownership, Household appliance ownership

(2012, 2013)). Therefore, according to these authors, the MEPI makes use of double limits and sets the threshold for each indicator minimum achievement. In addition, it performs the classification of households as deprived or non-deprived on the basis of each indicator. Thus, the following quantities and conditions are defined:

(x) is the observed variable's achievement for the $i - th$ household from all n population samples, on the jth variable from m observed variables, where $x_{ij} > 0$. Therefore: $X = \left[x_{ij}\right]_{n\times m}$ is a $n \times m$ status achievement matrix, t_j is the deprivation threshold on jth variable; $D = \left[d_{ij}\right]_{n\times m}$ is the ith household's multi-dimensional weighted-deprivation matrix in the jth variable. If $x_{ij} \leq t_j$ then $d_{ij} = w_j$, and if $x_{ij} > t_j$ then $d_{ij} = 0$. If $t_i = \sum m,\ j = 1d_{ij}$ is a multi-dimensional weighted deprivation column vector for the ith household.

The next step is to identify households with multi-dimensional weighted-deprivation in the column vector-t_i by defining limits: $(k > 0)$ and assuming energy-poor condition for households when the weighted deprivation exceeds k: $t_i(k) = 0$ if $t_i \leq k$ and $t_i(k) = t_i$ if $t_i > k$. Thus, $t(k)$ indicates the censored multi-dimensional energy deprivation average weighted vector.

The incidence (headcount ratio), which represents the number of households that are multi-dimensionally energy poor, is given as follows:

$$H = \frac{q}{n} \tag{2.9}$$

where: H is the incidence, q is the energy-poor people, and n is the number of sample households.

At this stage, the MEP intensity (A), which indicates the weighted censored multi-dimensional energy deprivation average:

$$A = \frac{\sum_{i=1}^{n} t_i(k)}{q} \tag{2.10}$$

Therefore, Finally, the MEPI (M) could be estimated using the following:

$$M = H \times A = \frac{q}{n} \times \frac{\sum_{i=1}^{n} t_i(k)}{q} \tag{2.11}$$

It could be observed that the MEPI metric provides an opportunity to measure the dimensions of energy poverty, which are in line with the concept of energy justice. In other words, to achieve energy justice, availability, affordability, accessibility, energy consumption as well as energy services must be adequate.

2.3 Current Strategies for Curbing Energy Poverty in SSA

The multiple metrics presented in the preceeding section are very useful for measuring the incidence of energy poverty in urban and/or rural households. Although, this acquired set of knowledge is not always accurate means to assess energy poverty condition in many cases, however, it should consist of essential step towards the development of appropriate strategies capable to alleviate or eradicate energy poverty. Therefore, several energy poverty curbing strategies have been proposed. These strategies could be grouped as follows.

2.3.1 Deployment of Renewable Energy

Nduka (2021) proposed the deployment of PV-based renewable energy resource to curb energy poverty in rural households of Nigeria. This author used the contingent valuation method (CVM) in order to determine responses and assess the willingness to pay (WTP) for a pico-photovoltaic (pico-PV) system in the context of improving access to clean energy. Furthermore, the lack of economic incentives, multiple and higher duties on RE technologies, lack of policy synergy and infrastructure problems are reported to be significant challenges faced by attempts to curb energy poverty in rural areas of Nigeria. However, this study indicates that rural households show strong preferences for sustainable energy, and the transition to clean energy would save each household 60 USD annually.

In an attempt to address concerns related to lack of incentives in the deployment of renewable energy, Batool et al. (2023) analysed the usefulness of corporate social responsibility (CSR) initiative in the mitigation of energy poverty in India. These authors made use of partial least squares structural equation modelling (PLS SEM) technique to evaluate the roles of the developed mediating variables (assessment of renewable energy resource, feasibility of sustainable energy supply, and sustainable energy development). Other than demonstrating the potential of renewable energy deployment in closing the gap between energy supply and demand in India, this study finds that CSR initiative is effective means to promote poverty alleviation in this part of the world.

2.3.2 Off-Grid and Mini-Grid Solutions

Due to the inherent limitations of conventional grid-based models in ensuring rural electricity access — particularly in remote areas — and growing concerns over climate change, there has been a global shift towards alternative energy solutions. Renewable energy-based off-grid systems have emerged as viable and increasingly popular options, providing sustainable and decentralized energy access to undeserved regions. According to BloombergNEF and SEforALL (2020), mini-grids consist of independent electrical networks powered by independent sources. These grids could operate either in stand-alone (off-grid) or in grid-connected mode. The following classification is usually applied: fossil-based mini-grids, renewable-based mini-grids, and hybrid renewable mini-grids. In an attempt to address the accessibility dimension of energy poverty in rural or remote areas, where there is no grid electricity, mini-grids are often relied upon. Domegni and Azouma (2022) indicated that about 11 million mini-grid connections have been made available in rural areas of SSA in the past decade. Paliwal et al. (2014) argued that the abundance of renewable energy sources (solar, etc.) in SSA justifies the attention afforded to this energy source as a major option for off-grid mini-grids in SSA. Therefore, among approximately 7,000 mini-grids in Asia and Africa, 63% are solar or solar hybrid systems, 21% hydro, 11% diesel/heavy fuel oil, and 3% biomass (Domegni and Azouma 2022).

2.3.3 Policy Support

Policy support consists of necessary strategy for curbing energy poverty in SSA. It fundamentally assists in setting clear and logical targets and indicators for access to energy, providing incentives and subsidies for RE-based decentralised solutions (off-grids and mini-grids, etc.), removing barriers and risks for private sector investment, strengthening regulatory frameworks, and promoting coordination and cooperation among various stakeholders. Nygaard et al. (2015) reported that administrative inefficiencies could unnecessary lengthen the time to secure regulatory license to 17 months. Babayomi and Okharedia (2019) highlighted key obstacles, including the absence of robust tariff-subsidy regulations and inadequate implementation of quality control measures, which hinder the expansion of mini-grids in SSA. To address these challenges and accelerate mini-grid adoption, these authors proposed targeted tariff-subsidy policies such as feed-in tariffs and feed-in premiums. These mechanisms can provide financial incentives to developers, ensure cost recovery, and enhance the financial viability of RE-based decentralised projects, thereby promoting wider energy access across the region.

2.3.4 Community Involvement

Engaging local communities in the planning, execution, and management of RE-based decentralised projects is a critical strategy for addressing energy poverty in SSA. Active community involvement ensures that energy solutions align with the needs and preferences of the population, enabling equitable distribution of benefits while minimizing potential social and environmental risks. Additionally, such participation fosters a sense of ownership, responsibility, and empowerment among energy users, enhancing their commitment to system upkeep and their willingness to contribute financially. This collaborative approach not only improves the long-term sustainability of these projects but also strengthens the social fabric and local capacity for development. Peters et al. (2019) argued that It is crucial to garner comprehensive community support by leveraging platforms and mechanisms that facilitate dialogue and collaboration. These may include mediation and consultation through trusted intermediaries such as community leaders, traditional rulers, and community development associations (CDAs). Additionally, engaging local government bodies and organizing inclusive town hall meetings can foster communal interaction and consensus-building. Such approaches help ensure that diverse voices are heard, promote transparency, and encourage collective ownership of initiatives, thereby enhancing the effectiveness and sustainability of development projects.

2.3.5 Innovative Business Models

According to Aziz et al. (2016), decentralised energy projects in SSA are predominantly financed through equity investments, grants, loans, and subsidies. These funding streams are sourced from a diverse array of stakeholders, including corporate entities, philanthropic foundations, and government agencies. However, this multifaceted financing approach often encounters challenges in achieving a balance between ensuring profitable business models that deliver returns on investment and addressing issues such as currency mismatches between capital cost and revenues, financial mismanagement, and unfavourable government policies. Overcoming these barriers requires strategic planning, robust financial frameworks, and innovative business model to enhance the viability and scalability of decentralized energy initiatives in the region.

2.4 Conclusion

This section reviews energy poverty assessment techniques, focusing on three widely used metrics: affordability, accessibility, and composite measures. A detailed discussion highlights that energy poverty in SSA is predominantly linked to the accessibility dimension, reflecting limited physical access to energy infrastructure. Furthermore, the energy poverty situation in SSA is closely tied to its socio-economic context and regulatory environment, suggesting that composite metrics, which integrate multiple dimensions, may offer the most comprehensive assessment approach. However, some proposed strategies for mitigating energy poverty involve qualitative or complex variables that are difficult to quantify, limiting their applicability within composite metrics. This underscores the inherent complexity of energy poverty assessment, as no single metric can universally capture all dimensions of energy deprivation across diverse contexts. A tailored, multi-faceted approach is therefore necessary for accurate evaluations and effective policy interventions.

References

Antepara, I., Papada, L., Gouveia, J.P., Katsoulakos, N., Kaliampakos, D.: Improving energy poverty measurement in southern European regions through equivalization of modeled energy costs. Sustainability **12**, 5721 (2020)

Ayodele, T.R., Ogunjuyigbe, A., Opebiyi, A.A.: Electrical energy poverty among micro-enterprises: indices estimation approach for the city of Ibadan. Nigeria. Sustain. Cities Soc. **37**, 344–357 (2018)

Aziz, S., Chowdhury, S.A., Groh, S.: The success of solar diesel minigrids in Bangladesh: A case study of sandwip island. Energy Proc. **103**, 316–321. https://doi.org/10.1016/J.EGYPRO.2016.11.292

Babayomi, O., Okharedia, T.: Challenges to Sub-saharan Africa's renewable microgrid expansion - a cetep solution model, in Proceeding of the 2019 IEEE PES/IAS Power Africa, Abuja, Nigeria (2019) (IEEE). 10.10.1109/PowerAfrica.2019.8928865

Barrella, R., Romero, J.C., Linares, J.I., Arenas, E., Asín, M., Centeno, E.: The dark side of energy poverty: Who is underconsuming in Spain and why? Energy Res. Soc. Sci. **86**, 102428 (2022)

Batool, K., Zhao, Z., Irfan, M., Żywiołek, J.: Assessing the role of sustainable strategies in alleviating energy poverty: an environmental sustainability paradigm. Environ. Sci. Pollut. Res. **30**, 67109–67130 (2023). https://doi.org/10.1007/s11356-023-27076-0

Bennett, M., Cooke, D., Price, C.W.: Left out in the cold? New energy tariffs, low-income households and the fuel poor. Fisc. Stud. **23**, 167–194 (2002)

Betto, F., Garengo, P., Lorenzoni, A.: A new measure of Italian hidden energy poverty. Energy Policy **138** (2020)

Bhatia, M., Angelou, N.: Beyond Connections: Energy Access Redefined. Tech. rep. Washington, D.C.: Energy Sector Management Assistance Program, World Bank Group, pp. 1–244 (2015)

BloombergNEF and SEforALL: State of the global mini-grids market report 2020: Trends of renewable energy hybrid mini-grids in Sub-saharan Africa, Asia and island nations. United Kingdom: BloombergNEF and Sustainable Energy for All. https://www.seforall.org/system/files/2020-06/MGP-2020-SEforALL.pdf

Boardman, B.: Fixing Fuel Poverty: Challenges and Solutions. London, UK, Earthscan (2010)

Boardman, B.: Fuel poverty synthesis: lessons learnt, actions needed. Energy Policy **49**, 143–148 (2012)

Boqiang, L., Okyere, M.A.: Multidimensional energy poverty and mental health: micro-level evidence from Ghana. Int. J. Environ. Res. Public Health **17**, 6726 (2020)

Bouzarovski, S., Tirado Heredo, S.: Geographies of injustice: the socio-spatial determinants of energy poverty in Poland, the Czech Republic and Hungary. Post-Communist. Econ. **29**(1), 27–50 (2017)

Bouzarovski, S., Petrova, S.: A global perspective on domestic energy deprivation: overcoming the energy poverty-fuel poverty binary. Energy Res. Soc. Sci. **10**, 31–40 (2015)

Bradshaw, J.: (2008) Who is fuel poor? Poverty. **131**, 9–11 (2008)

Castañ de la Rosa, R., Solís-Guzmán, J., Rubio-Bellido, C., Marrero, M.: Towards a multiple-indicator approach to EP in the European Union: a review. Energy Build. **193**, 36–48 (2019)

Cong, S., Nock, D., Qiu, Y.L., Xing, B.: Unveiling hidden energy poverty using the energy equity gap. Nat. Commun. **13**, 2456 (2022)

Critchley, R., Gilbertson, J., Grimsley, M., Green, G.: Warm front study group: living in cold homes after heating improvements: evidence from warm-front, England's Home Energy Efficiency Scheme. App, Energy **84**, 147–158 (2007)

Culver, L.C.: Energy Poverty: What You Measure Matters (2017). https://ngid9.sites.stanford.edu

Domegni, K., Azouma, Y.O.: Productive uses of energy: a solution for promoting energy justice in rural areas in West Africa. Renew. Sustain. Energy Rev. **160**, 112298 (2022). https://doi.org/10.1016/J.RSER.2022.112298

Drago, C., Gatto, A.: Gauging energy poverty in developing countries with a composite metric of electricity access. Utilities Policy **81**, 101486 (2023). https://doi.org/10.1016/j.jup.2022.101486

Dubois, U.: From targeting to implementation: The role of identification of fuel poor households. Energy Policy **49**, 107–115 (2012)

European Union (EU): Commission Recommendation (EU). 2020/1563 of October 14, 2020 (2020). https://www.interregeurope.eu/powerty/news/news-article/11957/recommendation-eu-2020-1563-energy-poverty

Fahmy, E. : The definition and measurement of fuel poverty. A briefing paper to inform consumer focus's submission to the Hills fuel poverty review , University of Bristol (2011)

Faiella, I., Lavecchia, L.: Energy poverty. How can you fight it, if you can't measure it?. Energy Build. 233 **11**(06) (2020)

Foster, V., Tre, J.P., Wodon, Q.: Energy prices, energy efficiency, and fuel poverty. Tech. rep. World Bank Group, pp. 1–7 (2000)

Goldemberg, J., Johansson, T.B.: Energy as an instrument for socio-economic development. Tech. rep. New York: United Nations Development Programme, pp. 1–80 (1995)

Goldemberg, J., T.B. Johansson, Reddy, A.K.N., Williams, R.H.: Energy for a Sustainable World. Tech. rep. Washington, D.C. World Resources Institute, pp. 1–73 (1987)

Gordon, D., Adelman, L., Ashworth, K., Bradshaw, J., Levitas, R., Middleton, S., Pantazis, C., Patsios, D., Payne, S., Townsend, P., Williams, J.: Poverty and social exclusion in Britain. York, Joseph Rowntree Foundation (2000)

Gordon, D., Mack, J., Lansley, S., Main, G., Nandy, S., Patsios, D., Pomati, M.: The Impoverishment of the UK - PSE UK first results: Living Standards. PSE, UK (2013)

Groh, S., Pachauri, S., Rao, N.D.: What are we measuring? An empirical analysis of household electricity access metrics in rural Bangladesh. Energy Sustain. Dev. **10**, 21–31 (2016)

Halkos, G., Kostakis, I.: Exploring the persistence and transience of energy poverty: evidence from a Greek household survey. Energ. Effi. **16**, 50 (2023). https://doi.org/10.1007/s12053-023-10137-1

Hartono, D., Hastuti, S.H., Balya, A.A., Pramono, W.: Modern energy consumption in Indonesia: assessment for accessibility and affordability, Energy. Sustain. Dev. **57**, 57–68 (2020). https://doi.org/10.1016/j.esd.2020.05.002

Healy, J.D.: Housing, Fuel Poverty and Health: A Pan-European Analysis. Ashgate, Aldershot (2004)

Healy, J.D., Clinch, J.P.: Fuel poverty, thermal comfort and occupancy: results of a national household-survey in Ireland. Appl. Energy **73**, 329–343 (2002)

Healy, J.D., Clinch, J.P.: Fuel poverty, thermal comfort and occupancy: results of a national household-survey in Ireland. Appl. Energy **73**, 329–343 (2002)

Hills, J.: Fuel poverty: The problem and its measurement. Interim report of the fuel poverty review. Centre for Analysis and Social Exclusion. Case Report 69, 192 p. (2011). ISSN 1465-3001

Hills, J.: Getting the measure of fuel poverty: Final report of the fuel poverty review. Case Report, 72, 237 p. (2012). ISSN 1465-3001

International Energy Agency (IEA): Defining energy access:2020 methodology (2020). https://www.iea.org/articles/defining-energy-access-2020-methodology

Ismail, Z., Khembo, P.: Determinants of energy poverty in South Africa. J. Energy in Southern Africa (JESA) **26**(3) (2015)

Karpinska, L., Smiech, S.: Invisible energy poverty? Analysing housing costs in Central and Eastern Europe. Energy Res. Soc. Sci. 70 **10**(16) (2020)

Kaygusuz, K.: Energy services and energy poverty for sustainable rural development. Renew. Sustain. Energy Rev. **15**(2), 936–947 (2011). https://doi.org/10.1016/j.rser.2010.11.003

Khanna, R.A., Li, Y., Mhaisalkar, S., Kumar, M., Liang, L.J.: Comprehensive energy poverty index: measuring energy poverty and identifying micro-level solutions in South and Southeast Asia. Energy Policy **132**, 379–391 (2019)

Legendre, B., Ricci, O.: Measuring fuel poverty in France: which households are the most fuel vulnerable? Energy Econ. **49**, 620–628 (2015)

Malla, S.: Household energy consumption patterns and its environmental implications: assessment of energy access and poverty in Nepal. Energy Policy **61**, 990–1002 (2013)

Meyer, S., Laurence, H., Bart, D., Lucie, M., Kevin, M.: Capturing the multifaceted nature of energy poverty: lessons from Belgium. Energy Res. Soc. Sci. **40**, 273–283 (2018)

Mirza, B., Szirmai, A.: Towards a New Measurement of Energy Poverty: A Cross-Community Analysis of Rural Pakistan. Tech. rep. United Nations University, pp. 1–41 (2010)

Mould, R., Baker, K.J.: Documenting fuel poverty from the householders' perspective. Energy Res. Soc. Sci. **31**, 21–31 (2017)

Mullen, C., Wade, N.: Appliance Data and Multi-Tier Framework for household electrical load modelling. www.mecs.org.uk, Working Paper 03 Dec 2020. v.1.2

Nduka, E.: How to get rural households out of energy poverty in Nigeria: a contingent valuation. Energy Policy **149** (2021). https://doi.org/10.1016/j.enpol.2020.112072

Nussbaumer, P., Bazilian, M., Modi, V.: Measuring energy poverty: focusing on what matters. Renew. Sustain. Energy Rev. **16**(1), 231–243 (2012). https://doi.org/10.1016/j.rser.2011.07.150

Nussbaumer, P., Nerini, F., Onyeji, I., Howells, M.: Global insights based on the multidimensional energy poverty index (MEPI). Sustainability **5**(5), 2060–2076 (2013). https://doi.org/10.3390/su5052060

Nygaard, I., Hansen, U.E., Mackenzie, G., Pedersen, M.B.: Measures for diffusion of solar PV in selected African countries. Int. J. Sustain. Energy **36**(7), 707–721. https://doi.org/10.1080/14786451.2015.1086768

O'Sullivan, K.C., Howden-Chapman, P.L., Fougere, G.: Making the connection: the relationship between fuel poverty, electricity disconnection, and prepayment metering. Energy Policy **39**, 733–741 (2011)

Pachauri, S., Spreng, D.: Energy use and energy access in relation to poverty. Econ. Pol. Wkly **39**(3), 271–278 (2004)

Paliwal, P., Patidar, N.P., Nema, R.K.: Planning of grid integrated distributed generators: a review of technology, objectives and techniques. Renew. Sustain. Energy Rev. **40**, 557–570 (2014). https://doi.org/10.1016/J.RSER.2014.07.200

Papada, L., Kaliampakos, D.: Being forced to skimp on energy needs: A new look at energy poverty in Greece. Energy Res. Soc. Sci. **64** (2020)

Papada, L., Kaliampakos, D.: Measuring energy poverty in Greece. Energy Policy **94**, 157–165 (2016)

Peters, J., Sievert, M., Toman, M.A.: Rural electrification through mini-grids: challenges ahead. Energy Policy **132**, 27–31. https://doi.org/10.1016/J.ENPOL.2019.05.016

Qurat-ul-Ann, A.R., Mirza, F.M.: Meta-analysis of empirical evidence on energy poverty: the case of developing economies. Energy Pol. **141** (2020). https://doi.org/10.1016/j.enpol.2020.111444

Rademaekers, K., Yearwood, J., Ferreira, A., Pye, S., Hamilton, I., Agnolucci, P., Grover, D., Karásek, J., Anisimova, N.: Selecting indicators to measure energy poverty, European commission, DG energy (2016b). https://trinomics.eu/wpcontent/uploads/2016/06/Selecting-Indicators-to-Measure-Energy-Poverty.pdf. [Last access: May 2024]

Rademaekers, K., Yearwood, J., Ferreira, A., Pye, S., Hamilton, I., Agnolucci, P., Grover, D., Karásek, J., Anisimova, N.: Selecting Indicators to Measure Energy Poverty. Rotterdam, The Netherlands, Trinomics (2016)

Rizal, R.N., Hartono, D., Dartanto, T., Gultom, Y.M.L.: Multidimensional energy poverty: A study of its measurement, decomposition, and determinants in Indonesia. Heliyon **10**(3) (2024). ISSN 2405-8440, https://doi.org/10.1016/j.heliyon.2024.e24135

Scarpellini, S., Rivera-Torres, P., Suárez-Perales, I., Aranda-Usón, A.: Analysis of energy poverty intensity from the perspective of the regional administration: empirical evidence from households in southern Europe. Energy Policy **86**, 729–738 (2015)

Sen, A.: Development as Freedom. Oxford University Press, pp. 1–384 (1999)

Seuret-Jimenez, D., Robles-Bonilla, T., Cedano, K.G.: Measurement of energy access using fuzzy logic. Energies, MDPI **13**, 3266 (2020). https://doi.org/10.3390/en13123266

Siksnelyte-Butkiene, I., Streimikiene, D., Lekavicius, V., Balezentis, T.: Energy poverty indicators: a systematic literature review and comprehensive analysis of integrity. Sustainable Cities Soc. Elsevier **67**, 102756 (2021). https://doi.org/10.1016/j.scs.2021.102756

Sokołowski, J., Lewandowski, P., Kiełczewska, A., Bouzarovski, S.: A multidimensional index to measure energy poverty: the Polish case. Energy Sources Part B **15**(2), 92–112 (2020). https://doi.org/10.1080/15567249.2020.1742817

Tait, L.: Towards a multidimensional framework for measuring household energy access: application to South Africa. Energy Sustain. Dev. **38**(17), 1–9 (2017)

Thema, J., Vondung, F.: Expenditure-based indicators of energy poverty—an analysis of income and expenditure elasticities. Energies **14**(8) (2021)

Thomson, H.: Fuel Poverty Measurement in Europe: A Rapid Review of Existing Knowledge and Approaches Conducted for Eaga Charitable Trust. The University of York, Heslington, UK (2013)

Thomson, H., Bouzarovski, S., Snell, C.: Rethinking the measurement of energy poverty in Europe: a critical analysis of indicators and data. Indoor Built Environ. **26**(7), 879–901 (2017)

UN Sustainable Development Summit (SDS) : Sustainable development and disaster risk reduction (2015). http://www.un.org/sustainabledevelopment/summit

Waddams Price, C., Brazier, K., Pham, K., Mathieu, L., Wang, W.: Identifying Fuel Poverty Using Objective and Subjective Measures. Centre for Competition Policy Working Paper 07-11. Norwich: University of East Anglia (2007)

Chapter 3
Operational Concepts of Energy Systems in SSA

3.1 Introduction

The operational concepts of energy systems in SSA are pivotal in addressing the region's unique energy challenges and harnessing its abundant renewable energy resources. This vast and diverse region, home to over a billion people, faces significant energy access deficits, with many rural and remote areas lacking reliable electricity. Traditional energy sources, such as biomass, remain predominant, leading to environmental degradation and severe health concerns due to indoor air pollution. However, the region's rich solar, wind, hydro, and geothermal resources offer immense potential for sustainable energy solutions. Understanding and implementing effective operational concepts in energy systems is essential for optimizing these resources, improving energy access, and driving socio-economic development (IRENA and AfDB 2022).

This involves a multifaceted approach that includes integrating advanced technologies like smart grids and energy storage solutions, which can enhance the efficiency and reliability of energy supply. Upgrading and expanding energy infrastructure, particularly in underserved rural areas, is crucial for bridging the energy access gap. Additionally, fostering regulatory reforms that create a favourable environment for private sector investment and innovation is essential for the sustainable development of the energy sector (Agoundedemba et al. 2023).

Promoting investments in both centralized energy projects, such as large-scale solar farms and hydropower plants, and decentralized energy solutions, such as solar home systems and mini-grids, can address the diverse needs of the region. These decentralized solutions are particularly important for providing electricity to remote communities that are not connected to the national grid. Furthermore, capacity building and training programs for local technicians and engineers can ensure the effective operation and maintenance of these energy systems. By leveraging these operational concepts, SSA can transition towards a more resilient, inclusive, and sustainable energy future. This transition not only addresses the immediate energy needs of the

P. N. Bokoro and G. Sharma, *Curbing Energy Poverty in Sub-Saharan Africa*,
https://doi.org/10.1007/978-3-031-90072-3_3

population but also supports broader developmental goals, including poverty reduction, improved health outcomes, and environmental sustainability. The successful implementation of these concepts will require coordinated efforts from governments, international organizations, private sector stakeholders, and local communities (IEA 2022).

3.2 Energy Production in SSA

Energy production in SSA is a critical and dynamic sector that significantly influences the region's economic development and quality of life. Despite being rich in natural resources, including vast reserves of oil, gas, and renewable energy sources such as solar, wind, hydro, and geothermal, the region faces substantial energy challenges. A significant portion of the population lacks access to reliable electricity, particularly in rural and remote areas, resulting in heavy reliance on traditional biomass for cooking and heating. This reliance not only contributes to environmental degradation but also poses severe health risks due to indoor air pollution. Addressing these challenges requires a comprehensive approach that leverages the abundant renewable energy potential, enhances infrastructure, and fosters supportive policies and investments. As SSA seeks to expand its energy production capacity, the focus on sustainable and inclusive energy solutions becomes paramount. The region's future energy landscape will be shaped by its ability to effectively utilize its renewable resources, improve energy access, and support socio-economic development through innovative and resilient energy systems. The energy production in SSA by different sources from 2019 to 2024 is given in Table 3.1.

Table 3.1 Energy production in SSA by different sources of energy (TWh)

Year	Hydropower	Solar	Wind	Biomass	Natural gas	Coal	Oil
2019	60	5	3	15	40	30	25
2020	62	7	4	16	42	31	25
2021	65	10	5	17	44	32	24
2022	68	14	6	18	47	33	23
2023	70	18	7	19	50	34	22
2024	73	24	9	20	53	35	21

3.2.1 Conventional and Renewable Resources

SSA is endowed with both conventional and renewable energy resources, each playing a critical role in the region's energy landscape. The effective utilization of these resources is vital for addressing the widespread energy access challenges and supporting socio-economic development (Ideriah 2024). The energy contribution by different sources of energy in SSA is shown in Fig. 3.1.

1. Conventional Energy Resources: involve fossil fuels such as oil, gas, coal, etc. Nigeria, Angola, and Gabon possess significant oil and gas reserves. Nigeria is one of the largest oil producers in the region, with oil exports forming a major part of its economy. Similarly, Angola has vast offshore oil fields that contribute substantially to its GDP. Natural gas reserves, particularly in Mozambique and Tanzania, are also substantial, offering potential for energy production and export. South Africa possesses the largest coal reserves in the continent and relies heavily on this resource for electricity generation. Coal-fired power plants remain a significant part of the country's energy mix, providing a reliable source of power but also contributing to environmental concerns. The following challenges are associated with conventional resources:

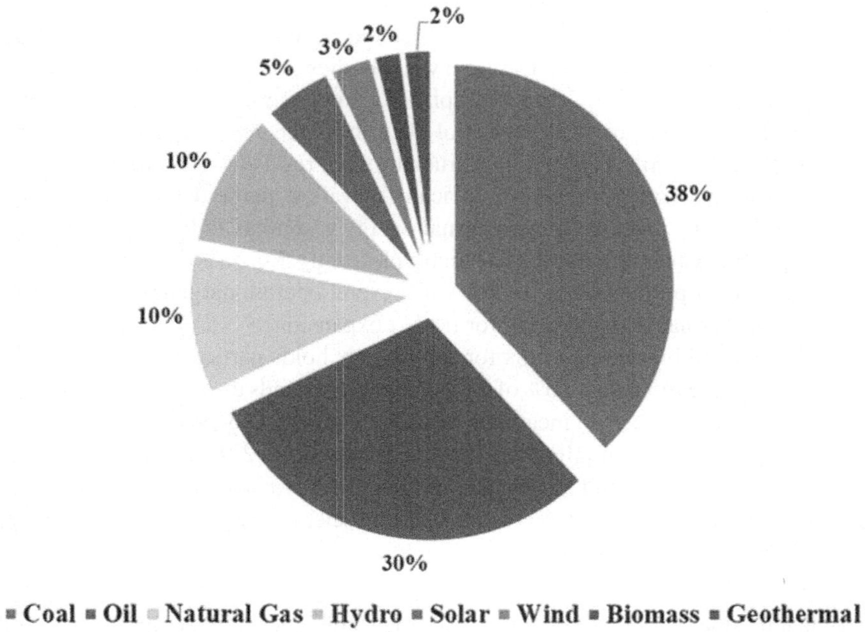

Fig. 3.1 Energy share by various sources of energy

a. Environmental Impact: the extraction and use of fossil fuels have significant environmental impacts, including air and water pollution, habitat destruction, and greenhouse gas emissions;
b. Economic Volatility: the dependence on fossil fuels can lead to economic instability due to fluctuating global oil prices, which can affect national revenues and economic planning.

2. Renewable Energy Resources: involve environmental-friendly resources such as:

a. Solar Energy: SSA enjoys high solar irradiance, making it one of the most promising regions for solar energy development. Countries like Kenya, South Africa, and Namibia are leading in the deployment of solar power projects (Bloom 2020), and it is commonly used for solar home systems, water heating, irrigation, powering small appliances, mini-grids, and utility-scale solar farms are increasingly being deployed to provide electricity to remote and underserved areas;
b. Wind Energy: coastal and highland areas, such as those in Kenya, Ethiopia, and South Africa, have significant wind energy potential for wind farms to be developed to harness this resource and contribute to the national grid, thus the launch of the lake Turkana wind power project in Kenya, which is one of the largest wind farms in Africa and illustrates the region's potential for large-scale wind energy generation;
c. Hydropower: several countries (DR Congo, Ethiopia, Zambia, etc.) in SSA region possess numerous rivers and water bodies suitable for large hydropower projects such as the Grand Ethiopian Renaissance and the Inga Dam, with the potential to provide substantial electricity generation. Small and micro-hydropower projects also play a crucial role in rural electrification;
d. Geothermal Energy: the East African rift valley, particularly in Kenya and Ethiopia, has substantial geothermal energy resources to build power plants capable to generate a stable and reliable electricity. Kenya is a leader in geothermal energy production in Africa, with several operational plants contributing to the national grid and plans for further expansion;
e. Biomass and Bioenergy: this form of energy holds particular importance for SSA, where more than 80% of the population depends on wood, crop residues, and animal residues to meet their household needs. This practice has environmental and health implications (World Bank Group 2010). There is growing interest in developing sustainable biomass supply chains and modern bioenergy technologies, such as biogas and biofuels, to provide cleaner and more efficient energy alternatives.

Balancing the use of conventional and renewable energy resources is essential for SSA's energy future. While conventional resources like oil, gas, and coal remain important, the shift towards renewable energy is crucial for sustainable development. By leveraging its abundant renewable resources, the region can improve energy access, reduce environmental impact, and enhance economic stability. This transition requires supportive policies, investments in infrastructure, and capacity building to ensure the effective deployment and management of diverse energy sources.

3.2.2 Management of Energy Production Infrastructure

SSA faces significant challenges in managing its energy production infrastructure. A combination of factors, including geographical, economic, and technological constraints, hinders the efficient and effective management of energy resources (Tomala et al. 2021). A number of steps involved in the management of infrastructure for energy production in SSA is displayed in Fig. 3.2.

1. Key Challenges:
   a. Underdeveloped Infrastructure: the region suffers from a lack of robust and reliable energy infrastructure. This includes limited transmission and distribution networks, which hampers the efficient delivery of electricity to consumers;
   b. Geographic Challenges: the vast and often remote nature of many areas in SSA makes infrastructure development and maintenance costly and difficult;

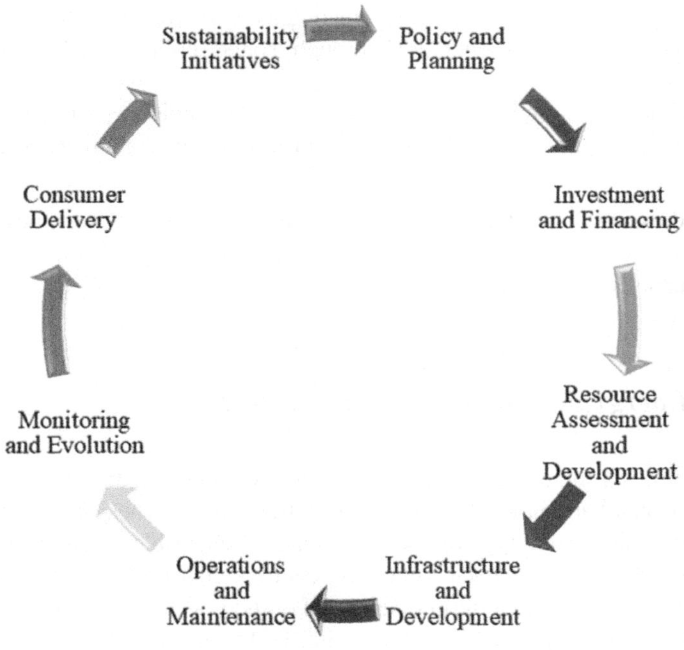

Fig. 3.2 Processes for the management of infrastructure for energy production

c. Financial Constraints: limited financial resources restrict investments in infrastructure development and upkeep, leading to aging and inefficient systems;
d. Technical Expertise: a shortage of skilled personnel with the necessary expertise to manage and maintain energy infrastructure is a persistent problem;
e. Governance and Corruption: weak governance structures and corruption can hinder efficient management and lead to losses and inefficiencies.

2. Strategies for Improvement: to address these challenges and improve the management of energy production infrastructure, several strategies can be implemented:

   a. Public-Private Partnerships: collaborating with private sector entities can bring in necessary capital, expertise, and efficiency to infrastructure development and management;
   b. Investment in Grid Expansion: expanding and upgrading transmission and distribution networks is crucial for improving energy access and reliability;
   c. Renewable Energy Integration: incorporating renewable energy sources into the grid requires careful planning and infrastructure upgrades to ensure stability and efficiency;
   d. Capacity Building: investing in training and development programs for energy sector professionals is essential to build a skilled workforce;
   e. Policy and Regulatory reforms: creating a conducive environment for investment and efficient operation through sound policies and regulations is vital;
   f. Technology Adoption: embracing innovative technologies, such as smart grids and digital platforms, can enhance infrastructure management and efficiency;

By addressing these challenges and implementing effective strategies, SSA can improve its energy production infrastructure, enhance energy access, and contribute to sustainable development.

3.2.3 Energy Production Investment Cost

Investment costs for energy production in SSA are a crucial factor in the region's efforts to expand and modernize its energy infrastructure. These costs vary significantly depending on the type of energy resource. Renewable energy projects, such as solar, wind, and hydropower, generally require higher initial capital investments but benefit from lower long-term operational costs. For instance, utility-scale solar projects typically range from 1,000 to 1,500 USD per kilowatt (kW), while wind projects cost between 1,500 and 2,500 USD per kW. Hydropower projects can vary widely, with small hydropower installations costing between 1,000 and 5,000 USD per kW and large-scale projects ranging from 1,000 to 3,000 USD per kW.

Conventional energy sources such as oil, gas, and coal also require substantial investments. The cost for oil and gas projects, particularly offshore ones, can range from 10,000 to 40,000 USD per barrel of oil equivalent per day (BOE/D) capacity, reflecting the complexity and scale of these operations. Coal-fired power plants, prevalent in countries like South Africa, have capital costs ranging from 1,000 to 3,500 USD per kW, including the expenses for construction, coal handling, and pollution control equipment. These conventional energy sources also entail higher operational and maintenance expenses compared to renewable energy projects (Moses 2023).

Several challenges further elevate the investment costs in SSA. Accessing affordable financing remains a significant hurdle due to high interest rates, perceived political and economic risks, and underdeveloped financial markets. Additionally, inconsistent and unclear policy frameworks can deter investors by introducing regulatory uncertainties and bureaucratic delays. Infrastructure and logistical challenges, especially in remote and rural areas, also contribute to higher project costs.

Innovative financing solutions are pivotal in addressing these challenges. Public-private partnerships (PPPs) can mobilize both public and private sector resources, sharing risks and reducing the financial burden on governments. Green bonds and climate finance initiatives attract investors interested in supporting sustainable and environmentally friendly projects. Support from multilateral development banks, such as the African Development Bank (AfDB) and the World Bank, provides crucial funding and technical assistance, helping to de-risk investments and mobilize private capital.

By improving regulatory environments and leveraging diverse funding mechanisms, SSA can attract the necessary investments to enhance energy access and promote sustainable economic development. Efforts to streamline regulatory processes, provide clear and supportive policy frameworks, and develop robust financial markets will be key in reducing investment costs and facilitating the growth of the region's energy infrastructure. This strategic approach will enable SSA to harness its abundant energy resources, ensuring a resilient and sustainable energy future for the region.

3.3 Grid Energy Systems

Grid energy systems in SSA play a fundamental role in the region's efforts to enhance energy access and support economic growth. The grid infrastructure, which includes generation, transmission, and distribution networks, is essential for delivering electricity to both urban and rural areas. However, the region faces significant challenges in grid energy systems, including limited coverage, frequent outages, and insufficient capacity to meet growing demand. Many areas, particularly rural and remote regions, remain underserved, relying on unreliable or non-existent grid connections. Despite these challenges, there are substantial opportunities for development. Expanding and modernizing grid energy systems, integrating renewable energy sources, and implementing advanced technologies such as smart grids and energy storage can significantly improve reliability and access. Additionally, regional cooperation and investments in infrastructure are crucial for creating a more resilient and efficient grid. By addressing these challenges and leveraging opportunities, SSA can build a robust grid energy system that supports sustainable development and enhances the quality of life for its population (Hugues 2024).

3.3.1 Reliability and Efficiency

The reliability and efficiency of grid energy systems in SSA are vital for fostering economic growth and improving the quality of life across the region. However, these systems encounter several significant challenges that hinder their effectiveness. Frequent power outages are a major issue, largely attributable to outdated infrastructure that is unable to handle the growing demands of the population and economy. Additionally, high technical losses occur due to inefficient transmission and distribution networks. These losses, often exceeding 20%, result from the aging equipment, poor maintenance, and inadequate infrastructure, which can lead to substantial energy wastage and reduced overall efficiency (Kaseke and Hosking 2013).

Grid coverage is another critical challenge. Many rural and remote areas are not connected to the grid, leaving a significant portion of the population without reliable electricity access. This uneven distribution exacerbates energy inequality and hampers development in underserved regions. To address these issues, substantial investment is needed to modernize and expand grid infrastructure. Upgrading transmission lines and replacing old equipment are crucial steps in reducing outages and technical losses. Furthermore, the adoption of smart grid technologies can greatly enhance grid reliability. Smart grids enable real-time monitoring and fault detection, which helps to quickly identify and address issues, reducing the frequency and duration of power outages (UNCTAD 2023).

Integrating efficient generation methods is also essential for improving the overall performance of grid systems. Transitioning from outdated and inefficient fossil fuel-based power plants to more modern and efficient technologies, such as combined cycle gas turbines, can enhance energy production efficiency. Additionally, incorporating renewable energy sources like solar, wind, and hydropower into the grid can diversify the energy mix, reduce dependency on fossil fuels, and provide a more sustainable and resilient energy supply.

Regional cooperation plays a crucial role in addressing these challenges. Initiatives such as the African Power Pool (APP) aim to facilitate cross-border electricity trading and collaborative infrastructure projects, which can enhance grid reliability and efficiency across countries. Supportive policies and regulatory frameworks are also essential for creating an environment that encourages investment in grid modernization and the adoption of advanced technologies. By fostering regional collaboration, implementing effective policies, and investing in modern infrastructure, SSA can significantly improve the reliability and efficiency of its grid energy systems, leading to more stable and effective electricity access and supporting the region's broader development goals.

3.3.2 Economic Impact and Sustainability

The economic impact and sustainability of grid energy systems in SSA are profound, influencing various aspects of development, economic growth, and environmental sustainability. Under the economic impact, the following factors are observed:

1. Growth and Development:

   a. Business Productivity: reliable grid energy systems are crucial for enhancing business productivity. Access to consistent electricity supports industrial activities, improves operational efficiency, and enables the growth of small and medium-sized enterprises (SMEs). For instance, businesses that experience frequent power outages face increased operational costs and reduced competitiveness;
   b. Job Creation: expanding and modernizing grid infrastructure creates numerous job opportunities in construction, maintenance, and energy management sectors. This employment generation contributes to economic development and poverty reduction.

2. Investment Attraction:

   a. Investor Confidence: a reliable and efficient grid energy system boosts investor confidence. Consistent electricity supply is a key factor for both local and international investors when considering investments in manufacturing, technology, and other sectors;
   b. Economic Stability: improved grid reliability reduces the economic risks associated with power outages and energy shortages, leading to more stable economic conditions and fostering a favorable environment for long-term investments.

3. Enhanced Services:

   a. Healthcare and Education: reliable electricity is essential for healthcare facilities and educational institutions. It ensures that hospitals can operate critical equipment and that schools can provide adequate learning environments, which in turn enhances the overall quality of public services and contributes to human capital development.

Under sustainability, the following factors are scrutinised:

1. Environmental Impact:

   a. Reduced Emissions: modernizing grid systems and integrating renewable energy sources significantly reduce greenhouse gas emissions. Renewable energy projects, such as solar, wind, and hydropower, provide cleaner alternatives to fossil fuels, contributing to global climate change mitigation efforts;
   b. Energy Efficiency: upgrading infrastructure to improve energy efficiency reduces wastage and optimizes the use of resources. Implementing smart grid technologies and efficient generation methods further supports sustainability by minimizing environmental impact.

2. Resource Management:

   a. Sustainable Resource Use: integrating renewable energy sources helps in the sustainable management of natural resources. Solar and wind energy rely on abundant natural phenomena that do not deplete finite resources, unlike fossil fuels;
   b. Diversification: diversifying the energy mix by incorporating various renewable sources reduces dependency on any single resource, enhancing the overall sustainability of the energy system.

3. Climate Resilience:

   a. Adaptation to Climate Change: modernized and diversified energy systems are more resilient to climate-related disruptions. For example, renewable energy sources can be less susceptible to the impacts of extreme weather events compared to conventional fossil fuel power plants;
   b. Local Benefits: renewable energy projects often involve local communities in their development and operation, providing additional benefits such as local job creation and enhanced community engagement, which contribute to long-term sustainability.

4. Economic Sustainability:

   a. Cost-Effectiveness: although the initial investment for renewable energy infrastructure can be high, the long-term operational and maintenance costs are generally lower compared to conventional fossil fuel plants. Over time, this leads to more sustainable economic benefits and reduced energy costs;
   b. Energy Security: investing in a diverse and resilient grid energy system enhances energy security by reducing reliance on imported fuels and mitigating the risks associated with global energy market fluctuations.

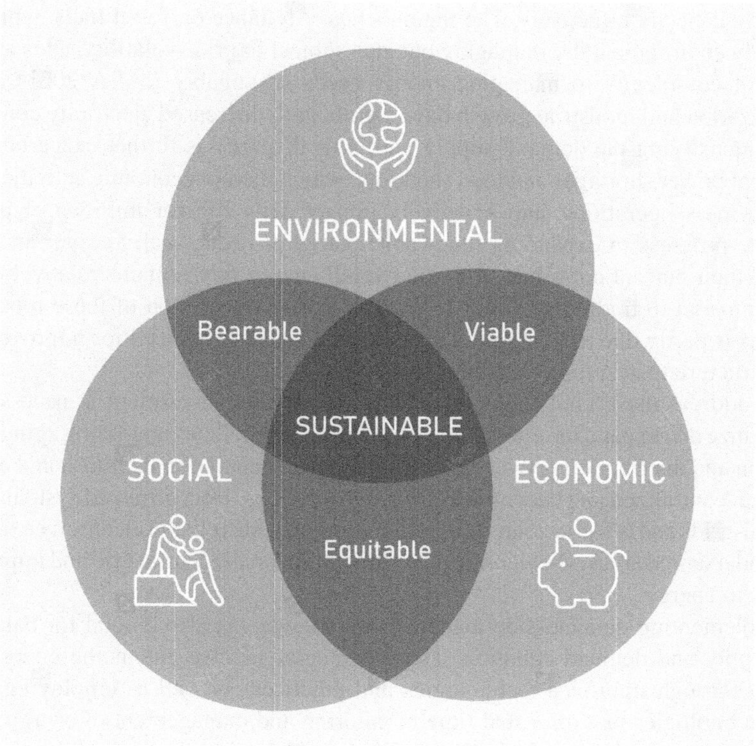

Fig. 3.3 Economic impact and sustainability in SSA

A block diagram is shown in Fig. 3.3 that illustrates the economic impact and sustainability in SSA. It highlights key sectors like agriculture, energy, infrastructure, trade, and education, and how they contribute to economic growth, environmental sustainability, and social development.

3.3.3 Capacity and Demand

The capacity and demand of grid energy systems in SSA present a multifaceted and challenging situation that profoundly affects the region's economic progress and overall quality of life. Currently, the energy capacity in many Sub-Saharan countries falls short of meeting the rising electricity demands due to several interconnected issues. Outdated infrastructure, characterized by aging power plants, inefficient transmission and distribution networks, and inadequate maintenance, contributes to a limited and unreliable energy supply. Furthermore, there has been limited investment in expanding and modernizing generation facilities, which hampers the ability to increase capacity effectively. The region's heavy reliance on fossil fuels, which are not only environmentally damaging but also subject to price volatility, adds another layer of complexity to managing energy needs sustainably (IAEA 2023). Rapid urbanization and industrial growth have significantly increased electricity consumption, intensifying the demand-supply gap. This disparity is further exacerbated by frequent power shortages and load shedding, which disrupt economic activities, hinder business operations, and negatively impact daily life for millions of people. Despite progress in expanding renewable energy projects, such as solar and wind farms, their current contribution to the overall energy mix remains relatively modest compared to traditional fossil fuels. The slower integration of these renewable sources is partly due to high initial investment costs and the need for improved grid infrastructure to accommodate their variability.

To address these challenges effectively, substantial investment is necessary to modernize and expand the energy infrastructure. This includes upgrading aging transmission and distribution networks, investing in new generation facilities, and exploring both centralized and decentralized energy solutions. Decentralized systems, such as mini-grids and off-grid solar installations, can provide reliable electricity to remote and underserved areas, alleviating some of the strain on the central grid and improving access to energy.

Implementing demand-side management strategies is also crucial for balancing the supply and demand equation. These strategies involve enhancing energy efficiency through improved technologies and practices, as well as employing smart grid technologies that offer real-time monitoring and management of energy flows. Smart grids can optimize electricity distribution, reduce losses, and integrate renewable energy sources more effectively. Regional integration initiatives, such as the African Power Pool (APP), offer additional opportunities to address capacity and demand issues. By facilitating cross-border electricity trading and shared infrastructure projects, these initiatives can help balance supply and demand more effectively and enhance overall grid stability. Ultimately, improving energy efficiency, adopting advanced technologies, and supporting robust policies and regulations are essential for enhancing grid capacity, better managing demand, and promoting sustainable economic growth in SSA. Through these efforts, the region can overcome current challenges and build a more reliable, efficient, and resilient energy system that supports its development goals and improves the quality of life for its population.

3.4 Off-Grid Energy Systems

Off-grid energy systems are becoming increasingly vital in SSA, addressing the significant challenge of providing reliable electricity to remote and underserved communities. In a region where traditional grid infrastructure is often insufficient or non-existent, off-grid solutions offer a practical and sustainable alternative. These systems, which include solar home systems, mini-grids, and portable solar devices, are tailored to meet the unique needs of dispersed populations. The off-grid energy

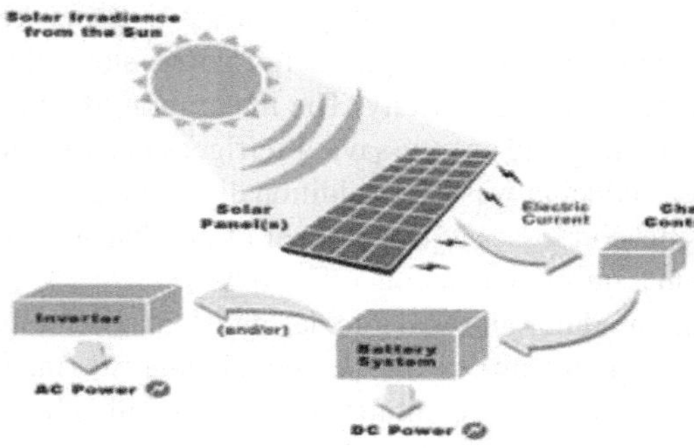

Fig. 3.4 Off-grid energy system in SSA

systems in SSA are illustrated in Fig. 3.4. It highlights renewable energy sources like solar, wind, hydropower, and biomass, showing the flow of energy from generation to storage, distribution, and end-users. The reliance on off-grid energy is driven by the region's rapid population growth, economic development, and the urgent need to improve living standards. Off-grid energy systems not only provide basic lighting and power for household appliances but also support critical services such as health-care, education, and small businesses. Additionally, they contribute to environmental sustainability by reducing dependence on fossil fuels and lowering greenhouse gas emissions. The adoption of off-grid energy technologies is supported by innovative financing models, such as pay-as-you-go schemes and microfinance, making them accessible to low-income households. As SSA continues to develop, off-grid energy systems are poised to play a crucial role in bridging the energy access gap and fostering inclusive, sustainable growth.

3.4.1 Reliability and Efficiency

The reliability and efficiency of off-grid energy systems are essential for addressing the energy needs of remote and underserved areas in SSA. These systems, particularly those based on solar energy, offer exceptional reliability due to several technological advancements. Durable photovoltaic panels are designed to withstand harsh environmental conditions, providing a consistent energy supply. Improved battery storage solutions, such as lithium-ion batteries, play a crucial role in maintaining stability by storing excess energy generated during sunny periods and ensuring power availability during the night or on cloudy days. The decentralized nature of off-grid systems significantly enhances their resilience. Unlike centralized grid systems, where a single failure can lead to widespread outages, each off-grid unit operates independently. This independence minimizes the risk of extensive power disruptions, ensuring a more reliable energy supply for remote communities. Moreover, the involvement of local communities in the operation and maintenance of these systems fosters better upkeep and reliability.

In terms of efficiency, off-grid energy systems are highly effective. Modern solar PV technology boasts high energy conversion rates, efficiently transforming sunlight into usable electricity. The integration of energy-efficient appliances, such as LED lighting and energy-saving refrigerators, maximizes the utility of the generated power, reducing wastage. Additionally, the local generation of electricity eliminates transmission losses associated with long-distance power delivery in traditional grid systems, further enhancing efficiency.

The modular and scalable design of off-grid solutions is another key advantage. These systems can be expanded incrementally to match growing energy needs, allowing households and communities to start with a basic setup and gradually add more capacity as required. This adaptability ensures that resources are used efficiently without the need for significant overhauls. Collectively, these attributes make off-grid energy systems a reliable and efficient solution for improving energy access in SSA. By providing a stable and efficient power supply, off-grid systems support sustainable development, enhance the quality of life, and drive economic growth in remote and underserved regions.

3.4.2 Economic Impact and Sustainability

The economic impact and sustainability of off-grid energy systems in SSA are profound, offering transformative benefits for the region's development while promoting environmental stewardship. Under economic impact, the following factors are observed:

1. Job Creation and Economic Growth:
   a. Local Employment: off-grid energy projects create job opportunities in installation, maintenance, and operation of systems, fostering local employment and skill development. The growth of small and medium-sized enterprises (SMEs) in the energy sector also stimulates economic activity. According to the International Renewable Energy Agency (IRENA), the renewable energy sector in SSA employed approximately 290,000 people in 2020, with a significant portion of these jobs in off-grid solar. The International Energy Agency (IEA) estimates that achieving universal energy access by 2030 could create up to 4.5 million direct and indirect jobs in the region, many of which would be related to off-grid and mini-grid solutions;
   b. Business Productivity: reliable access to electricity boosts the productivity of businesses, particularly in rural and remote areas. SMEs can extend their operating hours, improve product quality, and increase output, contributing to local economic growth.
2. Cost Savings:
   a. Reduced Energy Costs: off-grid systems can lower energy costs for households and businesses by providing a more affordable alternative to diesel generators

or other expensive and polluting energy sources. Solar energy, in particular, offers significant cost savings over time due to low operational and maintenance expenses. Off-grid solar home systems can save households up to 10–15 USD per month compared to kerosene lighting and diesel generators. This translates to annual savings of 120–180 USD per household. SMEs using solar energy instead of diesel generators can reduce their energy costs by 50–70%, significantly improving their profitability and sustainability;

b. Income Generation: access to electricity enables new income-generating activities, such as refrigeration for perishable goods, charging services for electronic devices, and the operation of electric machinery, enhancing livelihoods and economic stability.

3. Attraction of Investment:

    a. Increased Investments: the successful implementation of off-grid energy systems attracts investment from both domestic and international sources. Investors are drawn to the potential for sustainable returns and the growing market for renewable energy technologies in the region. The off-grid solar market in SSA attracted investments of over 1.5 billion of USD between 2015 and 2020, reflecting growing investor confidence in the sector;
    b. Public-Private Partnerships: governments and private sector entities increasingly collaborate on off-grid projects, leveraging combined resources to expand energy access and drive economic development. The market for off-grid solar products is projected to grow at a compound annual growth rate (CAGR) of over 10% from 2021 to 2030, indicating robust future investment potential.

Under sustainability, the following are debated:

1. Environmental Benefits:

    a. Reduction in GHGs Emissions: off-grid systems, particularly those based on renewable energy sources like solar and wind, contribute to significant reductions in greenhouse gas emissions compared to conventional fossil fuel-based energy. This helps mitigate climate change and reduces air pollution. Off-grid solar systems have the potential to reduce CO_2 emissions by up to 20 million metric tons annually by replacing kerosene lamps and diesel generators. This is equivalent to taking 4.3 million cars off the road each year;
    b. Conservation of Natural Resources: by reducing reliance on non-renewable energy sources, off-grid systems promote the sustainable use of natural resources. Solar and wind energy harness natural processes without depleting finite resources. By shifting to solar and other renewable off-grid systems, SSA can avoid the consumption of millions of liters of diesel and kerosene annually, conserving valuable natural resources.

2. Energy Independence and Security:
    a. Localized Energy Production: off-grid systems enhance energy security by reducing dependence on imported fuels and centralized power generation. Localized energy production ensures a more resilient and self-sufficient energy supply, less vulnerable to external disruptions. Off-grid renewable energy installations contributed to an increase of approximately 3.5 GW of installed capacity in SSA by the end of 2020, enhancing energy security and resilience. The adoption of off-grid solutions has contributed to increasing electrification rates in several countries. For example, in Kenya, off-grid solar systems have helped raise the rural electrification rate to over 75%, up from less than 10% a decade ago;
    b. Resilience to Climate Impacts: renewable off-grid systems are often more resilient to climate impacts such as extreme weather events. For example, solar panels can continue to operate after natural disasters, whereas centralized grid infrastructure may be more prone to damage.
3. Social Sustainability:
    a. Improved Quality of Life: access to reliable electricity improves the quality of life by enabling better healthcare, education, and communication services. Electrified health clinics can provide essential services like vaccine refrigeration, and schools can offer extended learning hours and digital education tools. Electrified health facilities using off-grid systems can save thousands of lives annually by providing reliable power for essential medical services. For instance, reliable refrigeration for vaccines could prevent up to 1 million child deaths annually across the region. Electrified schools can extend learning hours and improve educational outcomes for millions of children. Studies show that students in electrified schools have a 35% higher likelihood of passing national exams compared to those in non-electrified schools;
    b. Community Empowerment: off-grid projects often involve community participation and ownership, fostering a sense of empowerment and responsibility. Communities engaged in managing their energy resources tend to experience greater social cohesion and development outcomes.

The economic impact and sustainability of off-grid energy systems in SSA are substantial, offering a pathway to both immediate and long-term benefits. By creating jobs, reducing costs, and attracting investment, these systems drive economic growth and enhance livelihoods. Simultaneously, they promote environmental sustainability by reducing emissions, conserving resources, and fostering energy independence. Off-grid energy systems not only improve the quality of life for individuals and communities but also contribute to the broader goals of sustainable development and climate resilience in the region. Through continued investment and innovation, off-grid energy systems can play a pivotal role in shaping a sustainable and prosperous future for SSA.

3.4.3 *Capacity and Demand*

Off-grid energy systems in SSA are essential for providing electricity to remote regions where extending the main grid is impractical. These systems primarily rely on renewable energy sources such as solar, wind, and small hydropower. Solar photovoltaic (PV) systems dominate, with small household units ranging from 50–200 watts and larger community systems up to 10 kW. Mini-grids, which can range from a few kW to several hundred kW, serve multiple households and businesses, providing a more robust solution. Wind energy is less common due to variable wind conditions, while small hydropower systems, ranging from micro-hydro (up to 100 kW) to mini-hydro (up to 10 MW), are viable in regions with consistent water flow. Biomass and biogas systems, using organic waste, also contribute to the energy mix.

Demand for off-grid energy spans basic household needs, community services, and productive uses. Households primarily require electricity for lighting, phone charging, and small appliances, consuming 20–100 kWh per year. Community services, such as schools and clinics, need reliable power for lighting, refrigeration for vaccines, and educational tools, significantly increasing daily demand to several kWh per day. Productive uses, including small businesses, agricultural processing, and other income-generating activities, drive higher energy needs. Solar-powered water pumps are crucial for agriculture and clean water access, varying based on agricultural cycles and daily water needs. The sector faces financial, technical, and regulatory challenges. High upfront costs and the need for financing mechanisms, such as pay-as-you-go (PAYG) models, are critical for widespread adoption. PAYG models, which allow users to pay for energy in small, manageable increments, have been successful in making solar home systems more accessible. Technical challenges include ensuring system reliability and building local maintenance capacity to handle repairs and maintenance effectively. Supportive policies and quality assurance standards are essential to foster market growth, protect consumers, and ensure the reliability of off-grid systems. Current trends show increasing investment from international donors, development banks, and private investors. Technological advancements in solar PV efficiency, battery storage, and smart grid technology are making these systems more efficient and cost-effective. The scalability of mini-grids and solar home systems is improving, reaching more remote areas and communities. The integration of digital solutions, such as mobile payments and smart meters, enhances sustainability and system management, providing real-time data and facilitating remote monitoring and maintenance. Off-grid energy systems are pivotal for socio-economic development, improving quality of life and fostering economic activities in underserved regions of SSA. They provide essential services, support education and healthcare, and enable small businesses and agricultural activities, contributing to the overall development and empowerment of these communities.

3.5 Energy Access and Cost Evaluation in SSA

3.5.1 Energy Access

Energy access in SSA remains a significant challenge, with the region having the lowest electrification rates globally. According to the International Energy Agency (IEA), as of 2020, 775 million people, approximately 48% of the region's population, lack access to electricity (IEA World Energy Outlook 2015). This starkly contrasts with Asia, where only one country has similarly low electrification rates, as reported by the World Bank and the International Monetary Fund (IMF) (World Bank Report 2018). In fact, SSA's electricity access rate of 2.8% in 2018, as per IMF data, significantly trails South Asia's 7.1% (Corfee-Morlot et al. 2018). Despite abundant natural resources and economic potential, SSA's economic growth and development are hindered by widespread electricity deprivation. This challenge is exacerbated by rapid population growth projections, with the UN Department of Economic and Social Affairs (DESA) predicting a doubling of the region's population to two billion by 2050 (UN DESA 2020). Under current energy policies, the IEA (2020) forecast that nearly 90% of the global population without electricity will reside in SSA by 2030. Table 3.2 outlines population growth and electricity access trends.

1. Central Africa Power Pool: established in 2003, the Central African Power Pool (CAPP) is a regional energy organization comprising ten member countries. Its primary goal is to facilitate regional energy trading and integration, aligned with the Economic Community of Central African States (ECCAS) energy policy aimed at fostering socio-economic development across the region. Currently, CAPP serves a population of over 190 million people. The collective installed electricity generation capacity within the pool stands at approximately 7,500 MW. Despite this capacity, the region faces significant challenges in energy access, with an average electricity access rate of about 26% (Kenfack et al. 2021). The key details for each member country, including their main energy sources, generation capacities, and access rates are illustrated in Fig. 3.5. Hydroelectric power is the dominant energy source in the region, with an estimated potential capacity exceeding 650 TWh. This vast hydro capacity underscores the region's reliance on renewable energy resources. However, a substantial portion of the remaining

Table 3.2 Electricity access in 2019, Population growth between 2019 and 2050 in SSA Sub-Regions (IEA 2020)

Sub-Region	Pop. growth (%)	Total access (%)	Urban (%)	Rural (%)
Central	119	31.5	44	6
Eastern	95.5	48.2	79	35
Southern	70.5	83.5	67	20
Western	102	44.6	87	28

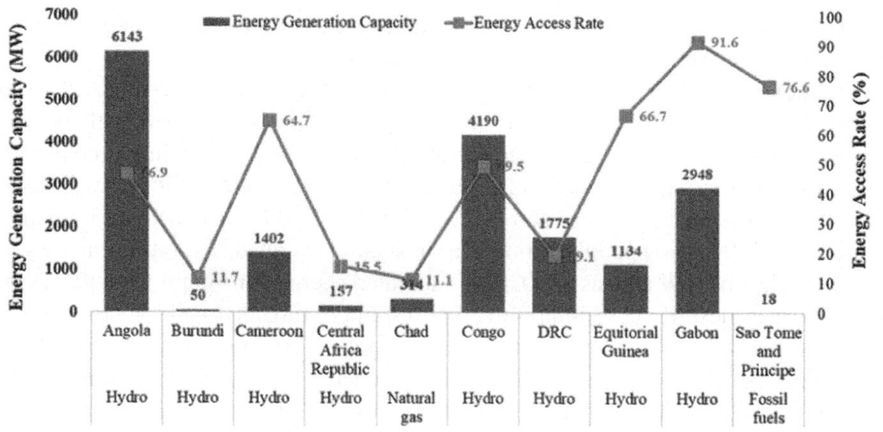

Fig. 3.5 Energy Statistics of CAPP member countries

energy capacity comes from thermal power plants, which contribute to the overall energy mix. Electricity access rates vary markedly among the member countries. Gabon has achieved the highest access rate in the pool, reflecting relatively better infrastructure and energy availability. In contrast, Chad, which relies predominantly on natural gas, faces the lowest access rate. This disparity highlights the uneven distribution of energy resources and infrastructure development within the region. Angola stands out with an electricity access rate of 46.9%, the highest in CAPP, and also possesses the largest energy generation capacity in the pool. This suggests a relatively more advanced energy infrastructure and better capacity to meet domestic demand compared to other member countries. Despite these resources and varying levels of access, CAPP encounters significant hurdles in fulfilling the region's energy needs. These challenges include:

a. Insufficient Incentives for Private Investment: the lack of effective policies and incentives to attract private sector investment in the energy sector hampers infrastructure development and capacity expansion;
b. Inadequate Power Infrastructure: many countries in the region suffer from underdeveloped power infrastructure, which limits their ability to generate, transmit, and distribute electricity efficiently;
c. High Power Losses: The region experiences considerable power losses due to outdated and inefficient infrastructure, reducing the overall effectiveness of the existing generation capacity;
d. Limited Grid Interconnection: the lack of sufficient grid interconnection between member countries restricts regional energy trading and cooperation, impeding the ability to optimize energy resources across borders.

These factors collectively contribute to CAPP's status as the least advanced among the four major regional power pools. Addressing these issues is crucial for enhancing regional energy integration, improving electricity access, and achieving sustainable socio-economic development across Central Africa.

2. East Africa Power Pool: in 2005, the Eastern African Power Pool (EAPP) was founded through the Inter-Governmental Declaration of Agreement, with the goal of improving regional energy cooperation among its eleven member countries. This initiative was designed to facilitate better interconnection of power systems and optimize energy trading across the region. The EAPP serves a large population of over 520 million people. It boasts a substantial installed generation capacity of more than 64 GW and, as of 2020, an estimated energy consumption of 491 TWh. However, despite these impressive figures, approximately 70% of the member countries suffer from low electricity access rates. This discrepancy highlights a significant challenge for the region, where energy access remains limited for a large portion of the population. The EAPP's primary objectives include enhancing the interconnection of power systems among member countries and promoting energy trading to achieve several key benefits. These benefits include optimizing energy usage, reducing the overall cost of power supply, and delivering social, economic, and environmental advantages (ICA 2016). The diverse energy profiles of the member countries, providing a clearer picture of their energy sources and infrastructure is illustrated in Fig. 3.6. Hydropower is the dominant energy source in the region, reflecting the abundant water resources available. This is followed by geothermal energy and fossil fuels such as oil and gas. Egypt stands out within the pool for having the highest energy generation capacity and the most advanced electricity access rate. On the other hand, Burundi has the lowest generation capacity, while South Sudan has the lowest electricity access rate, making it one of the most electrification-challenged countries as of 2020. The Eastern African region, despite its rich array of energy resources and strong economic growth potential, faces several barriers to progress. These challenges include:

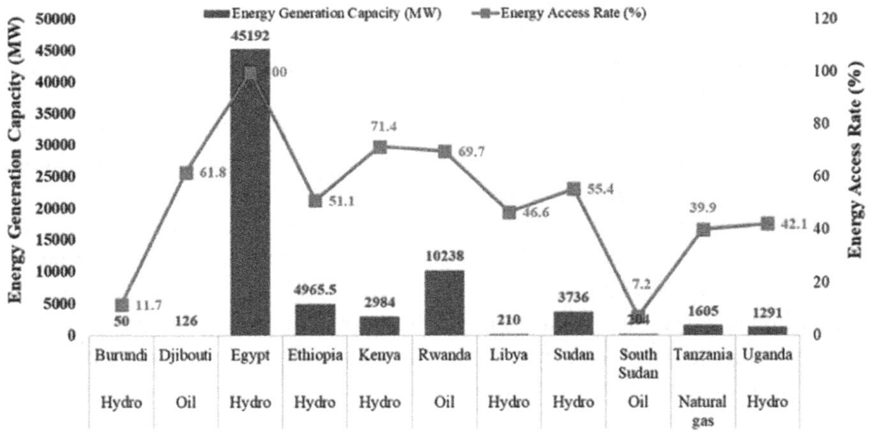

Fig. 3.6 Energy Statistics of EAPP member countries

a. Inadequate Infrastructure: many areas within the EAPP member countries lack the necessary infrastructure to support efficient energy generation, transmission, and distribution;
b. Insufficient Generation Capacity: the existing generation capacity is often inadequate to meet the growing energy demand, leading to frequent power shortages and unreliability;
c. Political Instability: political instability in various member countries can disrupt energy projects and deter investment, complicating efforts to improve energy infrastructure;
d. Limited Financial Resources: there is a shortage of financial resources required to invest in and develop the necessary energy infrastructure and technology;
e. Regulatory Barriers: regulatory obstacles can impede the development and integration of energy systems, hindering progress towards a more interconnected and efficient regional energy market.

The geographical positioning of the EAPP, with the Southern African Power Pool (SAPP) to the south and the Middle East to the north, underscores the importance of cross-border transmission, power interconnection, and regional energy trading. These factors are critical for improving energy access across the region. To overcome these challenges and foster development, the EAPP has outlined several strategic priorities. These include:

Strategic Priority 1: Increasing Generation Capacity: expanding and upgrading energy generation infrastructure to meet the growing demand and enhance reliability.

Strategic Priority 2: Investing in Technical Infrastructure: improving technical infrastructure such as transmission lines, substations, and grid systems to support efficient energy distribution and reduce losses.

Strategic Priority 3: Securing Financial Support: attracting and securing investments to fund energy projects and infrastructure development.

By addressing these strategic areas, the EAPP aims to enhance regional energy cooperation, improve electricity access, and support sustainable socio-economic development across Eastern Africa (Remy and Chattopadhyay 2020).

3. Southern Africa power pool: the Southern African Power Pool (SAPP), established in 1995, was designed to drive energy growth across the region through its targeted objectives. With a service area encompassing approximately 317 million people, SAPP is committed to enhancing its energy capacity to ensure a stable and affordable supply. This includes fostering energy trading among its member countries and advancing sustainable energy development through a balanced approach that addresses economic, environmental, and social factors (ECDPM 2019). SAPP is distinguished by its significant technological advancements. It has achieved an installed power capacity exceeding 71 GW and has an annual energy consumption

of about 353 TWh. The electricity access rate within the pool stands at 47%, making it the most technologically sophisticated power pool in Africa. This position is bolstered by its diverse energy portfolio, where coal and hydropower remain the primary sources of energy. However, SAPP is also making strides in incorporating alternative energy sources. Solar and wind power are being explored for their potential economic and environmental benefits (Monyei et al. 2022). The resource distribution across member countries varies significantly. Angola and Tanzania are notable for their abundant natural gas reserves, while South Africa is recognized for its substantial wind potential. Additionally, Mozambique, Swaziland, Zambia, and Tanzania each have considerable biomass resources. Furthermore, all member countries possess significant potential for solar energy, which could play a crucial role in future energy strategies. Despite these extensive resources, many member countries face challenges that limit their participation in the power pool. Issues such as inadequate energy generation capacity and low demand hinder their active engagement. The energy sources, generation capacities, and access rates for each member country within the pool are detailed in Fig. 3.7. South Africa is a key player in the regional energy market, holding approximately 51% of the installed generation capacity within SAPP. This substantial share allows South Africa to dominate the regional electricity trade. In contrast, many of the other member countries rely heavily on energy imports to meet their needs, highlighting the imbalance in energy distribution and the need for improved regional integration.

4. Western Africa Power Pool: founded in 1999 by the Economic Community of West African States (ECOWAS), the West African Power Pool (WAPP) is a regional energy initiative that includes fifteen member countries and serves a population exceeding 400 million. The primary goal of WAPP is to create a unified

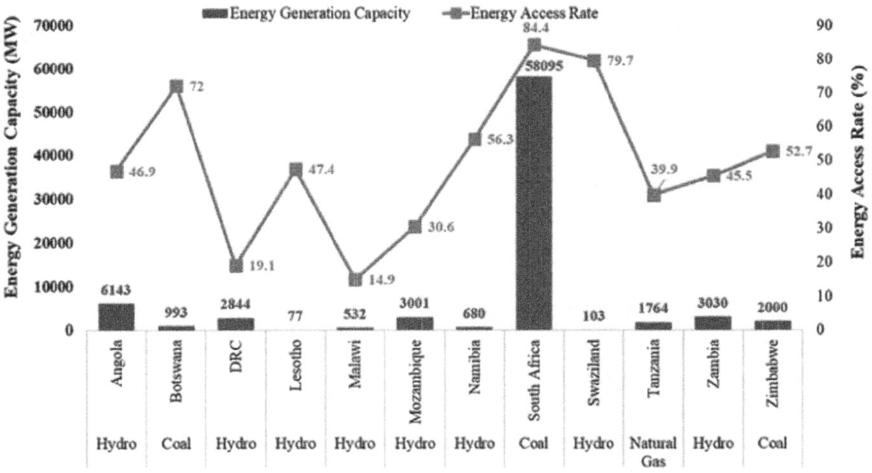

Fig. 3.7 Energy Statistics of SAPP member countries

energy market by integrating the national power networks of its member countries. This integration is intended to provide a reliable and secure power supply across the region. The pool has a total installed power capacity of approximately 23 GW. Despite this capacity, the region's annual energy consumption stands at around 58 TWh, and the electricity access rate is about 47% (Monyei et al. 2022). This indicates that while the installed capacity is substantial, a significant portion of the population still lacks reliable access to electricity. Nigeria, as the most populous member and a key resource-rich country, plays a pivotal role in WAPP. It contributes significantly with a generation capacity of 16,384 MW and an electricity access rate of 55.5%. Nigeria is also a major financial supporter of the pool, underscoring its critical position within the regional energy framework. A visual representation of the installed capacities and electricity access rates of each member country within the pool is provided in Fig. 3.8. The energy mix in the pool is dominated by natural gas, followed by hydropower, which constitutes about one-fifth of the total generation. Oil accounts for the remaining share of energy production. Nigeria and Ghana are significant producers of natural gas, which is used primarily in thermal power plants. This natural gas is also transported to neighbouring countries such as Benin, Niger, and Togo through the West African Gas Pipeline (IRENA 2015). Despite the region's considerable energy resources, eight of the member countries struggle with low electricity access rates and depend heavily on energy imports and diesel generators. This problem is aggravated by rising energy demand, leading to frequent power shortages and inadequate solutions to meet the needs of the population (ICA 2016). In response to these challenges, member countries are working to diversify their energy generation sources by incorporating renewable energy technologies, including solar and wind power. The regional policy sets a cap of 10% of peak demand for renewable energy generation (Smith et al. 2007). The renewable energy potential in the

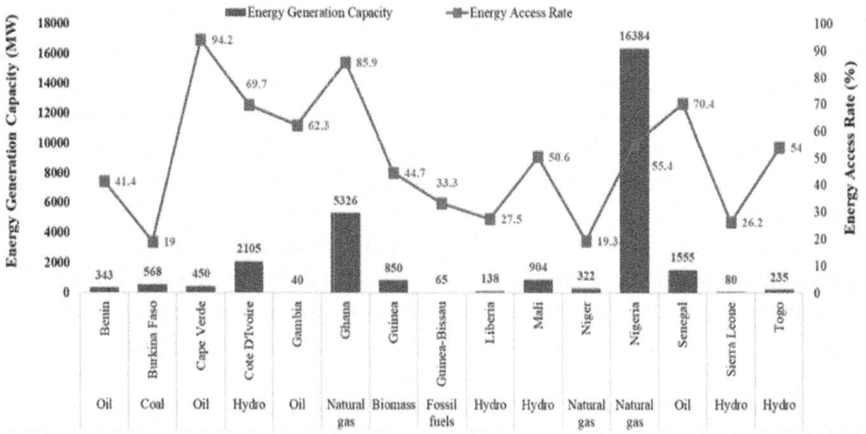

Fig. 3.8 Energy Statistics of WAPP member countries

region includes hydro, solar, and wind resources. Nigeria and Niger are particularly well-positioned to harness solar energy, with substantial potential available across most member countries, though Liberia and Mali have less potential in this area. For wind energy, Niger and Senegal have the highest quality resources, while Gambia, Nigeria, and Burkina Faso have moderate-quality wind resources. Additionally, many countries within the pool possess significant untapped hydroelectric potential. However, despite this considerable potential, the contribution of renewable energy — excluding hydro — to both centralized and decentralized generation remains minimal at present.

3.5.2 Energy Issues in SSA

Access to reliable, affordable, and sustainable energy is crucial for fostering economic development, improving living standards, and enhancing overall well-being. However, Africa is grappling with a range of energy-related challenges that hinder progress in these areas. These challenges include inadequate infrastructure, a lack of modern energy services, overreliance on fossil fuels, pervasive energy poverty, and significant environmental degradation. Energy poverty has long been a pressing issue in SSA. For decades, many countries in the region have struggled with low electricity access rates, often around 20%, and a shortage of modern energy facilities. The situation is exacerbated by rapid population growth, expanding economies, and increasing urbanization, all of which contribute to a surging demand for energy. This growing demand presents complex and multifaceted challenges that impact Africa's socio-economic development and its ability to meet sustainable development goals (Blimpo and Cosgrove-davies 2019). Yang and Yang (2018) argued that over 1.3 billion people worldwide still lack access to reliable energy, with nearly 61% of this population living in SSA. In response to these deficiencies, many African countries, including Nigeria, rely heavily on alternative energy sources such as firewood, coal, and petrol/diesel-powered generators (IEA 2022). While these alternatives provide some relief, they also introduce their own set of problems, such as depleting fossil fuel reserves, emitting greenhouse gases, causing environmental harm, and posing health risks. The energy grid infrastructure in the region further complicates access. Problems such as inefficiencies, outdated technology, insufficient financing, lack of maintenance, and limited generating capacity contribute to the challenges. Many energy sectors struggle with financial sustainability due to low revenues caused by poor billing practices, consumers' inability to afford energy, insufficient tariffs to cover operating costs, depreciation of energy investments, and poor management (ICA 2020). Additionally, the expansion of electricity grids to remote areas is hindered by high costs, losses in transmission and distribution, logistical issues related to inadequate road access, and unreliable electricity supply. Inadequate regulatory frameworks and political instability further exacerbate these challenges, preventing significant investment in the energy sector and stifling the adoption of new technologies and the growth of energy markets (Troost et al. 2018). Addressing these energy challenges requires substantial investments in modern energy services, integration of clean energy sources, and improvements in energy efficiency. By leveraging its abundant natural resources, Africa has the potential to develop its energy sector more effectively. This would not only help meet the growing energy demand but also support the region's efforts to achieve carbon reduction goals and promote sustainable development.

3.5.3 Regional Integrated Energy Systems in SSA

With a focus on energy as a key driver of economic growth and development, several countries in SSA are taking deliberate steps to overhaul their energy sectors. Recognizing the critical role that energy plays in fostering economic activities, these nations are prioritizing the integration of renewable energy resources, the adoption of energy-efficient technologies, and the creation of robust energy policies and strategies. These initiatives are aimed at transforming the energy landscape in SSA by diversifying energy sources, modernizing infrastructure, and streamlining licensing and market structures to create a more dynamic and sustainable energy sector. Despite the progress, the concept of Renewable Integrated Energy Systems (RIES) — which emphasizes the seamless integration of renewable energy sources into a cohesive and efficient energy system — has not yet gained the same level of traction in SSA as it has in developed countries like the United States and the United Kingdom, or in some rapidly developing regions with nearly universal energy access, such as China. However, the region is not without its advancements. Numerous projects, both existing and in the pipeline, are beginning to incorporate elements of RIES, reflecting a growing awareness and commitment to sustainable energy practices. One of the key trends in SSA's energy transition is the increasing focus on distributed generation. This approach, which involves producing electricity from various small-scale sources located close to where it is consumed, helps reduce transmission losses, improve grid stability, and bring power to remote or underserved areas. Alongside distributed generation, hybrid renewable energy systems — combinations of different renewable energy sources such as solar, wind, and hydropower — are becoming more common. These systems offer the flexibility to provide consistent energy supply by balancing the intermittency of individual renewable sources. Rural electrification is another crucial component of the energy transformation in SSA. With vast areas of the region still lacking access to reliable electricity, governments and private sector players are increasingly investing in off-grid and mini-grid solutions. These systems often leverage renewable energy and are seen as practical and sustainable ways to bring electricity to rural communities, where extending the traditional grid would be prohibitively expensive and technically challenging. Overall, while SSA still faces significant challenges in its energy sector, the ongoing efforts to incorporate renewable energy, improve infrastructure, and implement supportive policies indicate a positive trajectory. The adoption of RIES and related technologies, though still in its early stages, holds great promise for addressing the region's energy needs and contributing to its broader economic development goals.

1. Cameroon: this country has a population exceeding 25 million. It is the most densely populated country within the Central African Power Pool (CAPP). Despite its relatively high energy access rate of approximately 64% and a total energy generation capacity of 1,402 MW, the country struggles with significant energy insecurity (Kenfack et al. 2021). This paradox exists despite Cameroon's rich endowment of natural resources, including oil, gas, hydro, and solar energy (Muh et al. 2018). The root of this energy shortage can be traced to several intertwined factors. The country's rapidly growing population exerts increasing pressure on the existing energy infrastructure, while the effects of climate change further complicate energy production and distribution. Moreover, Cameroon faces chronic challenges such as insufficient funding for energy projects, a scarcity of private sector investment, and political instability, all of which hamper the development and expansion of the energy sector. In response to these challenges, the Cameroonian government has embarked on a series of initiatives aimed at reducing energy poverty and ensuring a more stable and sustainable energy supply. These initiatives include diversifying the energy mix to incorporate a greater proportion of renewable resources, investing in the installation of new power infrastructure, and enacting sustainable energy policies. Additionally, there is a concerted effort to foster collaboration between public and private sectors, recognizing that such partnerships are crucial for mobilizing the necessary capital and expertise. One of the key strategies employed to address the energy deficit has been the promotion of off-grid electrification solutions. These solutions are particularly important in rural and remote areas where grid extension is neither feasible nor cost-effective (Manjong et al. 2021). For instance, a study was conducted to evaluate the potential of Building Integrated Photovoltaic (BIPV) systems in Cameroon's tropical regions. The study analysed a scenario where a BIPV system was installed on the rooftop of an apartment, designed to meet a standard daily energy requirement of 3 kW. The results were promising, showing that the BIPV system could reduce annual energy consumption from 79.58 kWh/m to 13.64 kWh/m. Additionally, the system was found to lower building material and labour costs, making it an economically viable option for sustainable energy generation. In another effort to chart a sustainable energy future, a policy assessment was conducted using a back-casting methodology (Ayuketah et al. 2022). This assessment explored various scenarios for electricity generation in Cameroon, with the goal of determining the timeframe in which the country could significantly transition to renewable energy sources. The study provided valuable insights into the steps needed to shift towards a more renewable-centric energy system, helping to guide policy and investment decisions aimed at reducing the country's reliance on fossil fuels and enhancing energy security. These efforts reflect Cameroon's commitment to overcoming its energy challenges and achieving a more sustainable and reliable energy supply for its growing population. However, the success of these initiatives will depend on continued investment,

policy support, and the ability to navigate the complex socio-political landscape that influences the country's energy sector.

2. Kenya: following Nigeria and South Africa, Kenya ranks as the third-largest economy in SSA and the foremost economic powerhouse in East Africa, with a population exceeding 56 million people. The country's energy generation capacity surpasses 2.9 GW, with hydro and geothermal energy serving as the primary sources (Ang'u et al. 2020). Additionally, Kenya taps into biomass for biogas production and harnesses wind energy, showcasing significant potential, particularly in wind power, as highlighted in a review by (Takase et al. 2021). Despite its wealth of renewable energy resources, Kenya has grappled with persistent challenges in meeting its increasing energy demands. These challenges stem from various factors, including technical constraints, limited access to electricity in rural areas, escalating demand, inadequate funding for expanding generation capacity, and environmental concerns. Historically, Kenya has supplemented its energy needs by importing fossil fuels from neighbouring countries like Sudan. In recent years, however, Kenya has made substantial strides in enhancing energy access, establishing itself as a leader within the Eastern African Power Pool (EAPP). This progress is largely attributed to the creation of key energy agencies and the implementation of forward-looking policies. Organizations such as the Kenyan Electricity Generation Company, Kenya Electricity Transmission Company, Kenya Power and Lighting Company, and the Rural Electrification Authority have played pivotal roles in formulating national energy policies and developing a grid network aimed at ensuring the availability of affordable, reliable, and sustainable energy. Policies that have significantly contributed to the expansion of rural electrification include encouraging private ownership with minimal political interference, waiving tariffs, and exempting VAT on energy-related activities (Volkert and Klagge 2022). Kenya's concerted efforts to promote and diversify its renewable energy resources have also earned it a leading position in the off-grid solar market. However, many of the country's off-grid mini-grid systems remain diesel-based, raising environmental concerns due to the associated emissions. In response to these issues, a study by Moner-Girona et al. (2019) proposed a model to enhance energy access through decentralized rural electrification. This model involves the integration of traditional diesel generation with renewable sources such as solar, wind, and hydro energy, aiming to create a more sustainable and resilient energy system. One of the most notable renewable energy projects in Kenya is the Lake Turkana Wind Power development. As Africa's largest wind farm, this project is set to deliver 310 MW of reliable, clean, and low-cost energy to the national grid, significantly contributing to the country's renewable energy capacity and reducing its reliance on fossil fuels. This project exemplifies Kenya's commitment to leveraging its natural resources to meet its energy needs while fostering sustainable development (Gabisch et al. 2011).
3. South Africa: is located in the southern region of SSA, exemplifies a nation with both abundant energy resources and significant energy challenges. With a population exceeding 57.78 million people and a total energy generation capacity of 58 GW, South Africa has achieved an electricity access rate of over 80%

(Buraimoh et al. 2020). The country holds about 50% of the total energy capacity of the SSA grid, primarily relying on coal as its energy source. This heavy reliance on coal makes South Africa the largest emitter of greenhouse gases on the continent (John et al. 2016). Despite this, the country also possesses substantial solar energy resources and a more limited potential for wind energy, primarily concentrated along its coastal areas (Dagnachew et al. 2017). Although South Africa has the capacity to meet its energy demands through coal, the country has recently made significant strides toward aligning with the Paris Agreement on climate change. This shift has involved the development of new energy policies and substantial investments in renewable energy systems (Grobbelaar and Uriona 2018). These policies have been primarily formulated by the government through the state-owned power utility, Eskom, while the implementation of renewable energy projects has been largely driven by the private sector. The country's long-term energy plan aims at providing reliable and efficient energy services at competitive rates while targeting a 42% reduction in carbon emissions by 2025 (Ibrahim et al. 2021). The key policies supporting this transition include the National Development Plan (NDP), the Renewable Energy Independent Power Producer Procurement Program (REIPPPP), the Integrated Energy Plan (IEP), and the Integrated Resource Plan (IRP). These initiatives encompass both grid-connected and off-grid distributed generation systems, integrating available renewable energy resources. The availability of wind, solar, and hydropower resources across various parts of South Africa has facilitated increased rural electrification, the development of small and medium-scale distributed generation systems close to consumption centres, and the integration of these distributed generation systems into the national grid. While coal remains the backbone of South Africa's energy mix, the incorporation of wind and solar energy has helped to reduce carbon emissions, improve energy efficiency, and address the energy needs of local communities. However, the integration of renewable energy into the grid has been limited by several factors, including the uneven distribution of renewable energy resources and the lack of comprehensive research and studies to model and ensure the stability of a large, integrated energy system like the Renewable Integrated Energy System (RIES) (Thopil et al. 2018). In addition to these technical challenges, several other obstacles continue to impede the development of RIES in South Africa. These include the country's ongoing dependence on coal, the aging infrastructure of its energy system, and policy-related barriers such as the monopolization of the electricity utility, rising tariffs, bureaucratic red tape, high startup costs for renewable energy investments, and frequent load shedding (Hafner et al. 2018). In this context, a study conducted by Gebretsadik et al. (2016) focused on assessing the reliability of wind resources using optimal reservoir target power operations. This approach involves the integration of wind and hydropower systems to maximize energy generation and storage. The study found that this integration could increase wind penetration in South Africa's power system by 45%, alongside a corresponding reduction in coal power usage. Additionally, Akom et al. (2021) analysed South Africa's IEP and IRP, both of which support a comprehensive strategy for incorporating renewable energy sources into the grid. These policies are designed to ensure a reliable energy supply, optimize energy demand and efficiency, and evaluate economic benefits using metrics such as the net present value (NPV).

4. Nigeria: a detailed look at Renewable Integrated Energy Systems (RIES) in SSA, with a particular focus on Nigeria, reveals both the country's substantial energy potential and its ongoing challenges. Nigeria, as a member of the West African Power Pool (WAPP) and the most populous nation on the continent, is often referred to as the "Giant of Africa" due to its position as the largest economy in Africa. With a population of around 240 million, Nigeria is rich in both non-renewable and renewable energy resources, including oil, gas, hydro, solar, and wind energy. The nation's energy mix is heavily reliant on gas and hydropower, which together account for approximately 75% and 25% of its total energy supply, respectively (Diyoke et al. 2021). Despite its vast energy resources, Nigeria struggles with significant challenges in providing consistent and reliable electricity to its population. The country can only dispatch about one-third of its total energy capacity due to issues related to transmission infrastructure and operational constraints. Consequently, a substantial portion of the population, especially those in rural areas, lacks access to electricity, while urban areas suffer from frequent and unreliable power supply. Many Nigerians have resorted to alternative energy solutions such as diesel generators and solar systems for homes and small businesses (Adelaja 2020). Additionally, some have moved to urban centres in search of better, albeit still unreliable, electricity access (Ugwoke et al. 2021). The persistent energy crisis in Nigeria is driven by several factors, including inefficient utilization of available energy resources, inadequate planning and investment, outdated infrastructure, dependence on aging thermal power plants and large hydropower stations, and a deficiency in technical expertise for implementing advanced technologies like RIES (Okoro et al. 2021). In response, the Nigerian government has introduced various policies to diversify its energy resources and tackle the energy crisis. These include the National Energy Policy (NEP), the National Renewable Energy and Energy Efficiency Policy (NREEEP), the Sustainable Energy for All (SE4All) Action Agenda, the National Renewable Energy Action Plan (NREAP), and the Nigeria Energy Transition Plan (ETP) (Nwozor et al. 2021). Despite the ambitious nature of these policies, progress in integrating distributed generation systems into the national grid has been limited (Saka and Kiray 2020). Nevertheless, there has been a notable rise in the adoption of distributed energy solutions, particularly stand-alone or hybrid renewable systems like solar energy for households and small businesses. Mini-grids that combine technologies such as diesel generators, small hydropower, solar, and wind power have become increasingly common for rural electrification and local energy projects (Ugwoke et al. 2021). For example, Adaramola (2014) conducted research on the feasibility and economic viability of solar photovoltaic (PV) grid-tied systems for electricity generation in a specific region of Nigeria, using HOMER energy optimization software. Additionally, Dunmade (2016) investigated the potential and criteria for combining various renewable energy sources — such as solar, wind, and biomass — in Nigeria and assessed the sustainability of their integration.

3.5.4 Cost Evaluation

Cost evaluation for RIES in SSA involves a meticulous and multifaceted analysis aimed at ensuring the financial viability and operational sustainability of energy projects. Given the region's complex energy landscape, which features both abundant renewable resources and significant infrastructure challenges, a thorough cost evaluation is essential for the successful implementation of these projects. The process starts with an initial assessment to clearly define the project scope, identify specific energy needs, and evaluate the availability of renewable resources. This initial phase is crucial for estimating capital costs, which include expenses for equipment such as solar panels, wind turbines, and other renewable energy technologies, as well as costs associated with installation and regulatory permitting. Subsequently, operational costs are estimated, covering ongoing expenses for maintenance, fuel (in cases where hybrid systems are used), and labour. Additionally, decommissioning costs must be considered, which involve the removal and proper disposal of equipment at the end of its lifecycle. Financial modelling is a key component of the cost evaluation process. It involves conducting a cost-benefit analysis to determine the project's net present value (NPV) and return on investment. Sensitivity analysis is also performed to understand how fluctuations in critical variables, such as energy prices and interest rates, might impact the project's financial stability. Evaluating funding and financing options is another critical step, with possibilities including government grants, private investments, loans, equity financing, and public-private partnerships (PPPs). Risk assessment plays an integral role, addressing both financial and technical risks. Financial risks may involve currency fluctuations and political instability, while technical risks could include issues related to technology reliability and supply chain disruptions. Effective risk management strategies, such as currency hedging and developing robust supply chains, are essential for mitigating these risks (Panzer et al. 2020). The cost evaluation process concludes with a final review and approval of the budget, followed by the implementation and monitoring phase. During implementation, actual expenditures are monitored and compared against the budget, with financial models adjusted as needed to reflect real-world conditions. Overall, cost evaluation for RIES in SSA necessitates a detailed, dynamic approach to ensure that energy solutions are economically viable and capable of addressing the region's energy needs in a sustainable and efficient manner (Luhangala et al. 2022). The cost of providing energy in SSA varies significantly depending on the technology used, the scale of deployment, and regional factors. Here's a breakdown of the costs associated with different energy solutions:

1. Grid Extension: extending the national grid to rural and remote areas is often prohibitively expensive, with costs ranging from 10,000 to 20,000 USD per kilometre. Household connection fees can also be a barrier, with costs between 100 and 200 USD per household.
2. Off-Grid Solar Systems:

a. Solar Home Systems (SHS): these systems, which include solar panels, batteries, inverters, and lighting, range from 200 to 800 USD for a complete setup. Monthly payments for PAYG solar home systems typically range from 5 to 30 USD, depending on the system size and capacity.
b. Solar Lanterns: basic solar lanterns, which provide lighting and mobile phone charging, cost between 10 and 50 USD. These are widely used in rural areas due to their affordability and ease of use.

3. Mini-Grids: mini grids, which can provide electricity to entire communities, have an average cost of 1,200 to 2,500 USD per installed kW. The levelized cost of electricity (LCOE) for mini-grids ranges from 0.20 to 0.60 USD per kWh, which is competitive compared to diesel generators often used in remote areas. Despite the higher upfront costs, mini-grids offer more reliable and higher-capacity power than solar home systems, making them suitable for productive uses and community services.
4. Biomass and Biogas Systems: small-scale biomass and biogas systems, which convert organic waste into energy, can cost between 500 and 2,000 USD depending on the size and technology. These systems are particularly effective in agricultural regions, where organic waste is abundant and can be converted into a reliable energy source.

3.5.5 Economic and Social Implications

Improving energy access has profound economic and social benefits:

1. Economic Growth: access to electricity enables small businesses to operate more efficiently, extends business hours, and supports the development of new enterprises. Studies indicate that rural electrification can increase household incomes by 30–50% in some regions. Electrification supports agricultural productivity through irrigation, processing, and storage, reducing post-harvest losses and increasing food security.
2. Education: electrified schools can offer evening classes, use modern educational tools, and improve the overall learning environment. Access to electricity is linked to higher educational attainment and improved literacy rates.
3. Healthcare: electrified health facilities can provide better services, including refrigeration for vaccines, sterilization of medical equipment, and improved lighting for surgical procedures. Electrification is associated with a reduction in child mortality rates and improved maternal health outcomes.
4. Quality of Life: households with electricity have better lighting, access to information through radio and television, and reduced indoor air pollution from burning biomass, leading to better health outcomes.

Energy access in SSA remains a critical challenge, with significant disparities between urban and rural areas. While the cost of extending the grid is often prohibitive, off-grid solutions such as solar home systems, mini-grids, and biomass systems offer viable alternatives. These solutions are not only more affordable but also have the potential to transform lives by supporting economic growth, education, healthcare, and overall quality of life. Increased investment, supportive policies, and technological advancements are essential to scaling up these solutions and achieving universal energy access in the region (Table 3.3).

Table 3.3 Cost evaluation of various energy systems in SSA

Energy system	Cost range in USD	Cost details	LCOE	Key features
Grid Extension	10,000–20,000 per km	High costs due to infrastructure development; household connection fees: 100–200 USD	N/A	Suitable for urban areas; expensive in remote locations
Solar Home Systems	200–800 per system	Includes solar panels, batteries, inverters, and lighting	0.50–0.80 USD/kWh	Scalable; PAYG models available for affordability
Solar Lanterns	10–50 per unit	Provides basic lighting and phone charging	0.30–0.50 USD/kWh	Affordable; widely used in rural areas
Mini grids	1,200–2,500 per kWh installed	Includes installation of solar, wind, or hybrid systems; covers multiple households	0.20–0.60 USD/kWh	Reliable for communities; supports productive uses
Small Hydro	1,000–5,000 per kWh installed	Ranges from micro-hydro (up to 100 kW) to mini-hydro (up to 10 MW)	0.10–0.30 USD/kWh	Effective in areas with consistent water flow
Biomass systems	500–2,000 per system	Converts organic waste into energy; costs vary by size and technology	0.20–0.40 USD/kWh	Useful in agricultural areas; can utilize local waste resources

3.6 Conclusion

The operational concepts of energy systems in SSA encompass a range of strategies tailored to the region's unique challenges and opportunities. Extending the grid is ideal for densely populated urban centres where economies of scale can be achieved, but in remote and rural areas, it is prohibitively expensive and logistically challenging, with costs often exceeding 20,000 USD per kilometres. Consequently, off-grid solutions, including solar home systems, mini-grids, and biomass systems, present more feasible and cost-effective alternatives. Solar home systems, typically costing between 200 and 800 USD, and mini grids, with installation costs ranging from 1,200 to 2,500 USD per kW, have proven particularly effective in bridging the energy gap for households and small businesses. Hybrid systems that integrate multiple renewable sources, such as solar, wind, and biomass, further enhance the reliability and sustainability of energy supply. Financial models such as pay-as-you-go (PAYG) have made these solutions more accessible to low-income households by allowing users to pay for energy in small, manageable increments rather than incurring a large upfront cost. Technological advancements in battery storage and smart grid technology have also improved the efficiency and management of these systems, ensuring a more stable and reliable energy supply. Despite the progress, significant challenges remain, including high initial capital requirements, the need for robust regulatory frameworks to support market growth, and the development of local expertise for maintenance and repair of energy systems. Addressing these challenges through increased investment from international donors and private investors, the implementation of supportive policies by governments, and continued innovation in renewable energy technologies is crucial for achieving universal energy access. By doing so, the region can foster socio-economic development, improve the quality of life for millions of people, and promote sustainable growth.

References

Adaramola, M.S.: Viability of grid-connected solar PV energy system in Jos, Nigeria. Int. J. Electric. Power Energy Syst. **61**, 64–69 (2014). https://doi.org/https://doi.org/10.1016/j.ijepes.2014.03.015

Adelaja, A.O.: Barriers to national renewable energy policy adoption: insights from a case study of Nigeria. Energy Strat. Rev. **30**, 100519 (2020). https://doi.org/https://doi.org/10.1016/j.esr.2020.100519

Agoundedemba, M., Kim, C.K., Kim, H.-G.: Energy Status in Africa: Challenges, Progress and Sustainable Pathways. Energies **16**(23), 7708 (2023). https://doi.org/10.3390/en16237708

Akom, K., Shongwe, T., Joseph, M.K.: South Africa's integrated energy planning framework, 2015-2050. J. Energy S. Afr. **32**(1), 68–82 (2021). https://dx.doi.org/10.17159/2413-3051/2021/v32i1a8517

Ang'u, C., Muthama, N.J., Oludhe, C., Chitedze, I.: The role of diversity, reserve margin and system structure on retail electricity tariffs in Kenya. Heliyon **6**(8), e04626 (2020). https://doi.org/https://doi.org/10.1016/j.heliyon.2020.e04626

Ayuketah, Y., Gyamfi, S., Diawuo, F.A., Dagoumas, A.S.: Power generation expansion pathways: a policy analysis of the Cameroon power system. Energy Strat. Rev. **44**, 101004 (2022). https://doi.org/https://doi.org/10.1016/j.esr.2022.101004

Blimpo, M.P., Cosgrove-davies, M.: Electricity Access in Sub-saharan Africa. Africa Development Forum; World Bank (2019). https://openknowledge.worldbank.org/server/api/core/bitstreams/a6cfde8d-0224-5546-8062-5c2e69b741cb/content

Bloom, D.: The Long, Good Life Demographics and Economic Well-Being (2020). https://www.imf.org/external/pubs/ft/fandd/2020/03/pdf/fd0320.pdf

Buraimoh, E., Adebiyi, A.A., Ayamolowo, O.J., Davidson, I.E.: South African electricity supply system: the past, present and the future. In: IEEE PES/IAS PowerAfrica, pp. 1–5 (2020)

Corfee-Morlot, J., Parks, P., Ogunleye, J., Ayeni, F.: Achieving clean energy access in sub-Saharan Africa: a case study for the OECD, UN Environment. World Bank project (2018). https://www.greenfinanceplatform.org/research/achieving-clean-energy-access-sub-saharan-africa

Dagnachew, A.G., Lucas, P.L., Hof, A.F., Gernaat, D.E.H.J., de Boer, H.-S., van Vuuren, D.P.: The role of decentralized systems in providing universal electricity access in Sub-Saharan Africa–a model-based approach. Energy **139**, 184–195 (2017). https://doi.org/https://doi.org/10.1016/j.energy.2017.07.144

Diyoke, C., Ngwaka, U., Onah, T.O.: Comparative assessment of a hybrid of gas turbine and biomass power system for sustainable multi-generation in Nigeria. Sci. Afr. **13**, e00899 (2021). https://doi.org/https://doi.org/10.1016/j.sciaf.2021.e00899

Dunmade, I.: Hybridizing renewable energy systems in Nigeria: a contextual framework for their sustainability assessment. Eur. J. Eng. Technol. **4**(5) (2016)

Gabisch, M., Duru, U., Anvaripour, T., Turner, T., Negatu, G.: African development bank group project updated environmental and social impact assessment (2011)

Gebretsadik, Y., Fant, C., Strzepek, K., Arndt, C.: Optimized reservoir operation model of regional wind and hydro power integration case study: Zambezi Basin and South Africa. Appl. Energy **161**, 574–582 (2016). https://doi.org/https://doi.org/10.1016/j.apenergy.2015.09.077

Grobbelaar, S.S., Uriona, M.: Learning curves in the wind and solar sectors in South Africa. In: IEEE PES/IAS PowerAfrica, pp. 1–467 (2018)

Hafner, M., Tagliapietra, S., de Strasser, L.: Prospects for renewable energy in Africa. In: Energy in Africa. Springer Briefs in Energy. Springer, Cham (2018). https://doi.org/10.1007/978-3-319-92219-5-3

Hugues, B.: Opportunities and risks in the African grid of the future (2024). https://www.camco.fm/work

Ibrahim, I.D., Hamam, Y., Alayli, Y., Jamiru, T., Sadiku, E.R., Kupolati, W.K., Ndambuki, J.M., Eze, A.A.: A review on Africa energy supply through renewable energy production: Nigeria, Cameroon, Ghana and South Africa as a case study. Energy Strat. Rev. **38**, 100740 (2021). https://doi.org/https://doi.org/10.1016/j.esr.2021.100740

Ideriah, S.: Africa's unique energy future and the untold story of its transition (2024). https://www.slb.com/resource-library/insights-articles/africas-unique-energy-future-and-transition-story

IEA: Africa Energy Outlook 2022. IEA, Paris (2022). https://www.iea.org/reports/africa-energy-outlook-2022. (Licence: CC BY 4.0)

Infrastructure Consortium for Africa (ICA) Report: Infrastructure Financing Trends in Africa–2019-2020. https://www.icafrica.org/en/

Infrastructure Consortium for Africa (ICA): Updated Regional Power Status in Africa Power Pools (2016). https://www.icafrica.org/fileadmin/documents/Publications/Regional-Power-Pools-report-April17.pdf

International Atomic Energy Agency (IAEA): Meeting Africa's Growing Energy Needs in a Sustainable, Affordable and Efficient Way (2023). https://www.iaea.org/newscenter/news/meeting-africas-growing-energy-needs-in-a-sustainable-affordable-and-efficient-way

International Energy Agency (IEA) World Energy Outlook: Methodology for Energy Access Analysis (2015). https://www.iea.org/reports/world-energy-outlook-2015

International Energy Agency (IEA):World Energy Outlook (2020). https://www.iea.org/reports/world-energy-outlook-2020/overview-and-key-findings
International Energy Agency: Africa Energy Outlook (2022). https://www.iea.org/reports/africa-energy-outlook-2022
International Renewable Energy Agency (IRENA): Africa Power Sector: Planning and Prospects for Renewable Energy, vol. 44 (2015). https://www.irena.org/-/media/Files/IRENA/Agency/Publication/2015/IRENA-Africa-Power-Sector-synthesis-2015.pdf
IRENA and AfDB: Renewable Energy Market Analysis: Africa and its Regions. International Renewable Energy Agency and African Development Bank, Abu Dhabi and Abidjan (2022). www.irena.org/publications
John, T.M., Ucheaga, E.G., Olowo, O.O., Badejo, J.A., Atayero, A.A.: Towards building smart energy systems in sub-Saharan Africa: a conceptual analytics of electric power consumption. In: Future Technologies Conference (FTC), pp. 796–805. (2016)
Kaseke, N., Hosking, S.G.: Sub-Saharan Africa electricity supply inadequacy: implications. East. Afr. Soc. Sci. Res. Rev. **29**, 113–132 (2013). https://doi.org/10.1353/eas.2013.0009
Kenfack, J., Nzotcha, U., Voufo, J., Ngohe-Ekam, P.S., Nsangou, J.C., Bignom, B.: Cameroon's hydropower potential and development under the vision of Central Africa power pool (CAPP): a review. Renew. Sustain. Energy Rev. **151**, 111596 (2021). https://doi.org/10.1016/j.rser.2021.111596
Luhangala, D.L., Ambole, A., Musango, J.K., Ceschin, F., Dulo, S.: Energy price modeling in sub-Saharan Africa: a systematic literature review. Environ. Res. Infrastruct. Sustain. **2**(1), 15001 (2022). https://doi.org/10.1088/2634-4505/ac3fee
Manjong, N.B., Oyewo, A.S., Breyer, C.: Setting the pace for a sustainable energy transition in central Africa: the case of Cameroon. IEEE Access **9**, 145435–145458 (2021). https://doi.org/10.1109/ACCESS.2021.3121000
Moner-Girona, M., Bódis, K., Morrissey, J., Kougias, I., Hankins, M., Huld, T., Szabó, S.: Decentralized rural electrification in Kenya: speeding up universal energy access. Energy Sustain. Dev. **52**, 128–146 (2019). https://doi.org/https://doi.org/10.1016/j.esd.2019.07.009
Monyei, C. G.; Akpeji, K. O.; Oladeji, O.; Babatunde, O. M..; Aholu, O. C..; Adegoke, D. and Imafidon, J. O. : Regional cooperation for mitigating energy poverty in Sub-Saharan Africa: A context-based approach through the tripartite lenses of access, sufficiency, and mobility. Renewable and Sustainable Energy Reviews, 159, 112209. https://doi.org/https://doi.org/10.1016/j.rser.2022.112209 (2022)
Moses, O.: Who Finances Energy Projects in Africa? (2023). https://carnegieendowment.org/research/2023/11/who-finances-energy-projects-in-africa?lang=en
Muh, E., Amara, S., Tabet, F.: Sustainable energy policies in Cameroon: a holistic overview. Renew. Sustain. Energy Rev. **82**, 3420–3429 (2018). https://doi.org/https://doi.org/10.1016/j.rser.2017.10.049
Nwozor, A., Oshewolo, S., Owoeye, G., Okidu, O.: Nigeria's quest for alternative clean energy development: a cobweb of opportunities, pitfalls and multiple dilemmas. Energy Policy **149**, 112070 (2021). https://doi.org/https://doi.org/10.1016/j.enpol.2020.112070
Okoro, E.E., Adeleye, B.N., Okoye, L.U., Maxwell, O.: Gas flaring, ineffective utilization of energy resource and associated economic impact in Nigeria: evidence from ARDL and Bayer-Hanck cointegration techniques. Energy Policy **153**, 112260 (2021). https://doi.org/https://doi.org/10.1016/j.enpol.2021.112260
Panzer, A.D., Emerson, J.G., D'Cruz, B., Patel, A., Dabak, S., Isaranuwatchai, W., Teerawattananon, Y., Ollendorf, D.A., Neumann, P.J., Kim, D.D.: Growth and capacity for cost-effectiveness analysis in Africa. Health Econ. **29**(8), 945–954 (2020). https://doi.org/10.1002/hec.4029
Remy, T., Chattopadhyay, D.: Promoting better economics, renewables and CO2 reduction through trade: a case study for the Eastern Africa Power Pool. Energy Sustain. Dev. **57**, 81–97 (2020). https://doi.org/https://doi.org/10.1016/j.esd.2020.05.006
Saka, B., Kiray, V.: Distributed energy system in Nigeria: potentials, technologies, benefits and challenges. In: IEEE PES/IAS PowerAfrica, pp. 1–5 (2020)

Smith, J.C., Milligan, M.R., DeMeo, E.A., Parsons, B.: Utility wind integration and operating impact state of the art. IEEE Trans. Power Syst. **22**(3), 900–908 (2007). https://doi.org/10.1109/TPWRS.2007.901598

Takase, M., Kipkoech, R., Essandoh, P.K.: A comprehensive review of energy scenario and sustainable energy in Kenya. Fuel Commun. **7**, 100015 (2021). https://doi.org/https://doi.org/10.1016/j.jfueco.2021.100015

The European Centre for Development Policy Management (ECDPM): African power pools: regional energy. Natl Power (244), 1–34 (2019). https://ecdpm.org/work/african-power-pools-regional-energy-national-power (2019)

Thopil, M.S., Bansal, R.C., Zhang, L., Sharma, G.: A review of grid connected distributed generation using renewable energy sources in South Africa. Energy Strat. Rev. **21**, 88–97 (2018). https://doi.org/https://doi.org/10.1016/j.esr.2018.05.001

Tomala, J., Mierzejewski, M., Urbaniec, M., Martinez, S.: Towards Sustainable Energy Development in Sub-Saharan Africa: challenges and opportunities. Energies **14**(19), 6037 (2021). https://doi.org/10.3390/en14196037

Troost, P., Musango, J.K., Brent, A.C.: Strategic investment to increase access to finance among mini-grid ESCOs: perspectives from sub-Saharan Africa. In: 2018 2nd International Conference on Green Energy and Applications (ICGEA), pp. 29–37 (2018). https://api.semanticscholar.org/CorpusID:13714501

Ugwoke, B., Sulemanu, S., Corgnati, S.P., Leone, P., Pearce, J.M.: Demonstration of the integrated rural energy planning framework for sustainable energy development in low-income countries: case studies of rural communities in Nigeria. Renew. Sustain. Energy Rev. **144**, 110983 (2021). https://doi.org/https://doi.org/10.1016/j.rser.2021.110983

UN Department of Economic and Social Affairs (DESA): Global Policy Dialogue Series (2020). https://www.un.org/en/desa/policy-dialogue

UN trade & Development (UNCTAD): Improving energy access key to meeting development goals in Africa (2023). https://unctad.org/news/improving-energy-access-key-meeting-development-goals-africa

Volkert, M., Klagge, B.: Electrification and devolution in Kenya: opportunities and challenges. Energy for Sustainable Development, 71, 541–553. https://doi.org/https://doi.org/10.1016/j.esd.2022.10.022 (2022)

World Bank Group: Meeting Sub Saharan Africa's Energy Needs (2010). https://www.worldbank.org/en/news/feature/2010/12/22/biomass-meeting-sub-saharan-africas-energy-needs

World Bank Report (2018): Access to Energy is at the Heart of Development. Available on: https://www.worldbank.org/en/news/feature/2018/04/18/access-energy-sustainable-development-goal-7

Yang, F., Yang, M: Rural electrification in sub-Saharan Africa with innovative energy policy and new financing models. Mitig. Adapt. Strat. Global Change **23**(6), 933–952 (2018). https://doi.org/10.1007/s11027-017-9766-8

Chapter 4
Energy Markets and Regulatory Frameworks in SSA

4.1 Introduction

Energy markets and regulatory frameworks in SSA are both intricate and varied, mirroring the region's complex economic, social, and political conditions. The energy landscape in SSA has been shaped by a series of enduring challenges that have hindered the provision of reliable and affordable energy. These challenges stem from a combination of inadequate infrastructure, which includes aging and insufficient transmission and distribution networks, and limited access to financing, which has stymied both public and private investment in energy projects. The region's energy sector has traditionally relied heavily on biomass, particularly in rural areas where electricity access remains severely limited. This reliance on traditional energy sources has resulted in electricity access rates that are among the lowest globally, contributing to persistent energy poverty and hampering economic development. However, the energy sector in SSA has undergone significant changes in recent years. There is an increasing focus on energy sector reforms designed to improve operational efficiency, attract private sector investment, and expand access to modern energy services, especially in underserved rural and peri-urban areas. These reforms are part of a broader shift towards diversifying energy sources, particularly through the promotion of renewable energy. Many countries in the region have adopted policies aimed at incentivizing the development of renewable energy projects, recognizing the dual benefits of enhancing energy security and addressing environmental concerns (Luhangala et al. 2022).

Additionally, there has been a push towards encouraging off-grid and mini-grid solutions as a viable means of rapidly increasing energy access, particularly in remote areas where extending the national grid is economically unfeasible. To facilitate these reforms, regulatory frameworks are being updated to create a more enabling environment for private sector participation. This includes the introduction of competitive bidding processes for energy projects, tariff reforms, and the establishment of independent regulatory bodies to oversee the sector. Regional cooperation has

P. N. Bokoro and G. Sharma, *Curbing Energy Poverty in Sub-Saharan Africa*,
https://doi.org/10.1007/978-3-031-90072-3_4

also been a key focus, with initiatives such as the Southern African Power Pool (SAPP) and the West African Power Pool (WAPP) playing a critical role in enhancing cross-border electricity trade. These power pools aim to optimize the use of regional energy resources, reduce costs, and improve energy security by facilitating the sharing of electricity across borders. Despite these positive developments, the regulatory environment in many SSA countries continues to present significant challenges.

Regulatory uncertainty, often driven by frequent changes in policies and regulations, weak enforcement mechanisms, and limited capacity within regulatory agencies, remain substantial barriers to progress. These challenges are compounded by the need for significant capacity building within regulatory bodies to ensure that they can effectively oversee the sector and implement reforms (IRENA and AfDB 2022). Nonetheless, there is a growing recognition across the region of the critical importance of establishing robust and transparent regulatory frameworks. Such frameworks are essential not only for attracting the necessary investment to drive the development of sustainable energy markets but also for ensuring that these markets operate efficiently and equitably, providing affordable and reliable energy to all segments of the population (Eberhard et al. 2016).

4.2 Energy Markets in SSA

Energy markets in SSA are complex and rapidly evolving, characterized by significant diversity and substantial challenges. The region, home to over 1 billion people, faces a critical issue with electricity access, as nearly 600 million people, predominantly in rural areas, still lack basic electrical services (Luhangala et al. 2022). In these rural areas, electrification rates are often below 20%, severely limiting economic opportunities and quality of life. The energy market in SSA is predominantly reliant on traditional biomass, which makes up approximately 60% of the region's total energy consumption. This reliance on biomass is largely due to the limited availability of modern energy sources, particularly in rural and remote areas. The electricity generation mix in SSA varies significantly across different countries. Hydropower is a major source of electricity in several countries, particularly in Central and East Africa, where abundant water resources make it a viable option.

In contrast, other countries rely heavily on fossil fuels such as coal and natural gas, particularly in Southern and Western Africa. As of 2020, only about 48% of the population across SSA had access to electricity, highlighting the vast energy gap that still exists. Renewable energy, although growing, accounts for only about 22% of the region's total electricity generation, with hydropower being the dominant source (Copinschi 2022). To address these challenges, substantial investment is needed. It is estimated that the region requires approximately 40 billion annually to meet its energy access and sustainability goals by 2030. The dynamics of the energy market in SSA are characterized by a mix of state-owned utilities and an increasing number of independent power producers (IPPs). In recent years, there has been a significant uptick in private investment in renewable energy projects, particularly in countries like South Africa, Kenya, and Ethiopia. These countries have seen considerable growth in wind, solar, and geothermal power projects, driven by favourable policies and incentives for private sector participation (Dagnachew et al. 2023). Despite this progress, the energy market in SSA remains constrained by several challenges. Regulatory frameworks across the region are often inconsistent, with some countries making significant strides in liberalizing their energy markets, while others continue to struggle with outdated regulations, monopolistic practices, and a lack of transparency.

The establishment of competitive bidding processes for energy projects and the creation of independent regulatory bodies in some countries are positive steps, but these efforts are not yet widespread. Regional cooperation initiatives, such as the Southern African Power Pool (SAPP) and the West African Power Pool (WAPP), play a crucial role in enhancing energy trade and improving electricity reliability across borders. However, these initiatives also face challenges related to regulatory alignment, infrastructure development, and political stability (Sustainable Market Initiative's Energy Transition Task Force 2022). Overall, while the energy markets in SSA are evolving, significant hurdles remain. Addressing these challenges requires coordinated efforts to improve regulatory frameworks, increase investment, and enhance regional cooperation to ensure that the benefits of modern energy services are accessible to all.

4.2.1 Wholesale Energy Markets

Wholesale energy markets in SSA are in varying stages of development, reflecting the region's diverse economic and infrastructural conditions. In more developed markets, particularly in Southern Africa, efforts have been made to establish organized wholesale electricity markets, where electricity is traded between producers and large-scale consumers or distributors. The Southern African Power Pool (SAPP) is a leading example of this, facilitating cross-border electricity trade among its member countries. Established in 1995, SAPP is one of the few functioning regional power pools in Africa and serves as a platform for electricity trading among Southern African Development Community (SADC) member states. Despite these efforts, wholesale energy markets in SSA face several significant challenges. The region's energy infrastructure is often underdeveloped, with ageing transmission and distribution networks that are prone to inefficiencies and losses.

These infrastructural challenges are compounded by the high cost of energy generation and the financial instability of many state-owned utilities, which dominate the energy sector in most countries. As a result, the development of robust and competitive wholesale energy markets has been slow, with limited participation from independent power producers (IPPs) and private sector entities (IEA 2020). One of the key barriers to the development of wholesale energy markets in SSA is the lack of regulatory and institutional frameworks that support competitive market mechanisms. In many countries, electricity markets are still heavily regulated, with prices often set by governments rather than being determined by market forces. This regulatory environment can discourage investment in the energy sector, particularly from private companies that require clear and stable market conditions to operate profitably.

Another challenge is the limited capacity for cross-border energy trade outside of a few regions like Southern Africa. While the West African Power Pool (WAPP) and the East African Power Pool (EAPP) have been established to promote regional energy trade, these initiatives are still in the early stages of development and face numerous hurdles, including political and regulatory differences between member countries, inadequate transmission infrastructure, and limited financial resources (US EIA 2020). Despite these challenges, there are opportunities for growth in wholesale energy markets in SSA. The increasing interest in renewable energy, particularly solar and wind, presents a significant opportunity for the region. These resources are abundant and can be developed relatively quickly, providing a potential boost to the capacity of wholesale markets. Additionally, ongoing reforms aimed at liberalizing energy markets, improving regulatory frameworks, and enhancing regional cooperation are critical steps toward creating more dynamic and competitive wholesale energy markets. Wholesale energy markets in SSA are still in their nascent stages, there is significant potential for development. Achieving this potential will require addressing the region's infrastructural, regulatory, and financial challenges, as well as leveraging the growing interest in renewable energy and regional cooperation initiatives.

4.2.2 Retail Energy Markets

Retail energy markets in SSA are characterized by a mix of challenges and opportunities, shaped by the region's economic, infrastructural, and regulatory landscapes. The retail energy sector in the region typically involves the sale of electricity to end-users, including households, businesses, and industries. However, access to reliable and affordable electricity remains a significant issue, particularly in rural and peri-urban areas, where grid connectivity is often limited or non-existent. This has led to a complex and diverse market structure, with both grid-connected and off-grid solutions playing vital roles in meeting energy demand (Kambanda 2017).

In many countries, the retail energy market is dominated by state-owned utilities, which are responsible for the generation, transmission, and distribution of electricity. These utilities often operate as monopolies, with regulated tariffs set by government agencies. However, many state-owned utilities face financial difficulties, often resulting in inadequate investment in infrastructure, poor service quality, and frequent power outages. The financial instability of these utilities is often exacerbated by issues such as non-payment of bills, electricity theft, and inefficient billing systems. In response to these challenges, there has been a growing trend towards liberalization and the introduction of private sector participation in the retail energy market. In some countries, reforms have been implemented to allow independent power producers (IPPs) and private distributors to enter the market, particularly in the renewable energy sector. These reforms aim to increase competition, improve service delivery, and expand access to electricity. For example, in Kenya and South Africa, private companies have been increasingly involved in the distribution of electricity, particularly in areas where the national grid is not available (IRENA 2024). Off-grid and mini-grid solutions have also become an important component of the retail energy market in SSA, particularly in remote and rural areas.

These solutions often involve decentralized energy systems, such as solar home systems, mini-grids, and solar lanterns, which provide electricity to households and small businesses. The off-grid market has seen significant growth in recent years, driven by declining costs of solar technology, innovative business models, and support from international donors and development finance institutions. Companies like M-KOPA in Kenya and BBOXX in Rwanda have been successful in providing pay-as-you-go solar systems, which allow consumers to purchase electricity in small, affordable increments (WHO et al. 2023).

Despite these developments, the retail energy market in SSA faces several ongoing challenges. Regulatory frameworks in many countries are still underdeveloped, with issues such as lack of clarity on licensing procedures, inconsistent tariff policies, and weak enforcement of contracts. Additionally, the affordability of electricity remains a critical issue, particularly for low-income households. High tariffs, combined with the high upfront cost of off-grid systems, can make it difficult for many consumers to access and afford modern energy services. Furthermore, the reliability of electricity supply continues to be a major concern. Frequent power outages, voltage fluctuations, and poor-quality electricity are common issues that affect both grid-connected and off-grid consumers. These reliability issues not only impact the quality of life for households but also hinder the growth of businesses and industries, limiting economic development in the region. Overall, while there have been positive developments in the retail energy market in SSA, significant challenges remain. Addressing these challenges requires continued efforts to improve regulatory frameworks, increase investment in infrastructure, and expand access to affordable and reliable electricity for all consumers. The success of the retail energy market will be crucial for achieving broader economic and social development goals across the region.

4.2.3 Market Participants

The energy market in SSA involves a variety of participants, each playing a crucial role in the generation, distribution, and consumption of electricity. The structure of market participants varies across the region, influenced by the level of market development, regulatory frameworks, and the degree of private sector involvement. The main market participants include:

1. State-Owned Utilities (SOUs): according to Prasad (2011), SOUs are the dominant players in most SSA countries, responsible for electricity generation, transmission, and distribution. These entities often operate as vertically integrated monopolies, controlling all stages of the electricity supply chain. A few examples are Eskom in South Africa, SNEL in the DR Congo, KPLC in Kenya, and the Ghana Grid Company (GRIDCo) in Ghana. Many state-owned utilities face financial challenges, including debt, inefficiencies, and inadequate infrastructure, which can lead to unreliable service and frequent power outages.
2. Independent Power Producers (IPPs): are private companies that generate electricity and sell it to the national grid or directly to large consumers. According to Streatfeild (2018), IPPs play a critical role in increasing generation capacity, particularly in renewable energy sectors like solar, wind, and geothermal. Lake Turkana Wind Power in Kenya, Azura-Edo IPP in Nigeria, and the Gigawatt Global solar project in Rwanda could be listed among the few examples of these entities. IPPs have been instrumental in diversifying the energy mix and bringing in much-needed investment, helping to alleviate some of the supply constraints faced by state utilities.
3. Transmission System Operators (TSOs): according to Igbinovia and Tlusty (2014), TSOs are responsible for the high-voltage transmission of electricity from generation facilities to distribution networks. In many countries, TSOs are either state-owned entities or divisions within state-owned utilities. For example: in Ghana the TSO constitute a division of the power utility (GRIDCo). TSOs play a critical role in ensuring the stability and reliability of the electricity grid, managing cross-border electricity flows, especially within regional power pools.
4. Distribution Companies (DisCos): manage the delivery of electricity from the transmission grid to end consumers, including households, businesses, and industries. These companies may be state-owned, privately owned, or operate under public-private partnerships (USAID. gov/Power Africa 2018). Umeme in Uganda, and Nigeria's various DisCos like Eko Electricity Distribution Company and Ikeja Electric could be counted among the various DisCos in SSA. DisCos often struggle with issues such as electricity theft, non-payment of bills, and technical and commercial losses, which can impact their financial viability and the quality of service they provide.
5. Off-Grid and Mini-Grid Operators: these operators provide electricity to areas that are not connected to the national grid, often using renewable energy sources such as solar. They are particularly important in rural and remote areas where extending the grid is not economically viable (World Bank Group 2020). M-KOPA Solar in Kenya, BBOXX in Rwanda, and ZOLA Electric in Tanzania are a few examples of such entities. Off-grid and mini-grid solutions are vital for improving energy access in SSA, offering an alternative to traditional grid expansion.
6. Regional Power Pools (RPPs): according to Blyden and Davidson (2005), RPPs facilitate cross-border electricity trade, helping to optimize the use of energy resources and improve energy security among member countries. The SSA consists of four power pools: the Southern African Power Pool (SAPP), the West African Power Pool (WAPP), the Central African Power Pool (CAPP), and East

African Power Pool (EAPP). These pools enable countries to share surplus electricity, reduce costs, and enhance grid stability. However, they also face challenges related to infrastructure, regulatory harmonization, and political stability.
7. Regulatory Authorities: these bodies are responsible for overseeing the energy sector, setting tariffs, licensing market participants, and ensuring compliance with regulations. They play a crucial role in creating a conducive environment for investment and competition. L' Agence de Régulation de l' Electricité (ARE) in DR Congo, the Energy Regulatory Commission (ERC) in Kenya, the Nigerian Electricity Regulatory Commission (NERC), and the National Energy Regulator of South Africa (NERSA). These authorities often face capacity constraints and political pressures, which can affect their ability to enforce regulations effectively and maintain market stability.
8. Consumers: include residential, commercial, and industrial users who purchase electricity from utilities or private providers. In many parts of SSA, consumers face challenges such as high tariffs, unreliable supply, and limited access to modern energy services. Consumer demand, especially from growing urban populations and expanding industries, drives the need for increased generation capacity and improved service delivery.
9. Investors and Donors: include private equity firms, development finance institutions, and international donors. These institutions provide the capital needed for energy infrastructure projects. They are crucial for funding new generation capacity, especially in renewable energy, and supporting the expansion of off-grid solutions. The African Development Bank (AfDB), the World Bank, and private equity firms like Actis and Helios could be named among these institutions. These entities help bridge the financing gap in the energy sector, enabling the development of projects that might otherwise be unfeasible due to high costs or risks.

The energy market in SSA is shaped by a diverse group of participants, each with a specific role in the generation, transmission, distribution, and consumption of electricity. While the market faces significant challenges, particularly in terms of infrastructure, regulatory frameworks, and financial sustainability, the involvement of a wide range of stakeholders, including private sector participants, regional power pools, and international investors, offers opportunities for growth and development in the region's energy sector.

4.2.4 Pricing Mechanisms

Pricing mechanisms in SSA energy markets are shaped by a range of factors, including government policies, the structure of the energy sector, and the economic realities of the region (Oxford Institute for Energy Studies 2017). These mechanisms vary significantly across countries and are influenced by whether the energy sector is regulated by the government, partially liberalized, or fully open to market forces. An overview of the key pricing mechanisms in the SSA region is presented:

1. Cost-Plus Pricing:
   Cost-plus pricing is a widely utilized pricing mechanism in SSA, particularly in countries where state-owned utilities hold a dominant position in the energy sector. This model operates on the principle that electricity tariffs should reflect the total costs incurred by utilities in generating, transmitting, and distributing electricity to consumers. These costs include not only the direct expenses associated with fuel, labor, and maintenance but also a margin designed to cover operational overheads and fund necessary investments in infrastructure. In theory, cost-plus pricing aims to ensure that utilities can recover their expenses while maintaining a certain level of profitability to sustain operations and invest in the future. According to the OECD (2023), the practical application of this model in SSA is fraught with challenges, leading to several unintended consequences. One of the primary issues with cost-plus pricing in this region is the inefficiency that characterizes many state-owned utilities. These inefficiencies can manifest in various forms, such as overstaffing, where the number of employees far exceeds what is necessary for efficient operation, leading to inflated labor costs. Technical losses are another significant problem, with outdated and poorly maintained infrastructure resulting in a substantial portion of generated electricity being lost during transmission and distribution. Additionally, many of these utilities operate with aging infrastructure that is costly to maintain and inefficient in its energy use, further driving up operational costs. As a result of these inefficiencies, the cost of providing electricity becomes disproportionately high, which is then reflected in the tariffs charged to consumers. These high tariffs can place a considerable financial burden on households and businesses, particularly in a region where income levels are generally low. Ironically, despite these elevated tariffs, many state-owned utilities still struggle with financial sustainability. The revenues generated through these tariffs often fall short of covering the full spectrum of costs, including necessary investments in infrastructure upgrades and expansion. This shortfall can lead to a vicious cycle where utilities are unable to invest in improving their services, perpetuating inefficiencies and leading to further financial difficulties. Moreover, the financial deficits experienced by these utilities can have broader implications for the energy sector and the economy as a whole. Utilities that operate at a loss may require government subsidies to remain afloat, which can strain public finances and divert resources away from other critical areas of development. Additionally, the lack of financial viability can deter private investment in the sector, as investors may perceive the market as too risky or unprofitable.

2. Regulated Tariffs:
   In many SSA countries, electricity prices are regulated by government agencies or independent regulatory bodies, which play a crucial role in determining tariffs. These regulatory entities set electricity tariffs based on multiple considerations, including the actual costs of production, the necessity to attract investment in the energy sector, and social factors like ensuring affordability for low-income consumers. The balance between these factors is delicate, as setting tariffs too high can place a financial strain on consumers, while setting them too low can jeopardize the financial sustainability of utilities. For example, in Kenya, the Energy Regulatory Commission (ERC) is responsible for setting electricity tariffs. The ERC must consider the costs incurred by energy producers and distributors, while also ensuring that electricity remains affordable for the general population. Similarly, in South Africa, the National Energy Regulator of South Africa (NERSA) oversees the approval of tariff increases proposed by utilities such as Eskom, the country's largest power supplier. NERSA's role includes balancing the need for Eskom to cover its operational costs and fund new investments with the broader economic and social impacts of higher electricity prices on consumers. However, regulated tariffs can become a politically sensitive issue, especially in regions where the majority of the population may struggle with high living costs. In such cases, there is often pressure on regulatory bodies or governments to keep electricity prices artificially low to avoid public backlash. While this approach can help maintain short-term social stability. However, Klug et al. (2022) indicated that it often leads to significant challenges for the energy sector. When tariffs are set below the actual cost of producing and distributing electricity, utilities may struggle to recover their expenses. This financial shortfall can result in underinvestment in critical infrastructure, such as upgrading aging power plants, expanding transmission networks, and maintaining distribution systems. Over time, this underinvestment can degrade the quality of service, leading to more frequent power outages, reduced reliability, and slower progress in expanding electricity access to underserved areas (IMF Annual Report 2024). Moreover, the inability to recover costs can discourage private sector investment in the energy sector. Investors are less likely to finance new projects if they perceive the regulated tariffs as insufficient to provide a reasonable return on investment. This can stifle the development of new energy projects, particularly in renewable energy, where upfront capital costs are often high.
3. Subsidized Pricing:
   To make electricity more affordable, many governments in SSA subsidize energy prices. Subsidies can take the form of direct financial support to utilities, cross-subsidies where industrial users pay higher tariffs to subsidize residential users, or price caps on certain consumer categories. For instance, Nigeria has implemented cross-subsidies, where higher tariffs for commercial users help subsidize lower rates for residential consumers. Similarly, Ghana has provided direct subsidies to lower tariffs for consumers. While subsidies can improve affordability, they

often lead to significant fiscal burdens on governments. Additionally, they can distort market signals, discouraging investment in the energy sector and leading to inefficiencies.

4. Feed-in Tariffs (FiTs):
   According to Ndiritu and Engola (2020), FITs are used to encourage investment in renewable energy by guaranteeing a fixed price for electricity generated from renewable sources. These tariffs are usually higher than the market rate to make renewable energy projects financially viable. Several countries, including Kenya, Uganda, and Tanzania, have implemented FiTs to promote the development of renewable energy, particularly in wind, solar, and small hydro projects. This pricing system has been successful in attracting investment in renewables, but they can also lead to higher electricity prices for consumers if not managed carefully. Ensuring that FiTs are set at levels that reflect the decreasing costs of renewable technologies is crucial for long-term sustainability.
5. Time-of-Use (ToU) Pricing:
   ToU pricing is a dynamic pricing mechanism where electricity rates vary depending on the time of day, with higher prices during peak demand periods and lower prices during off-peak times. According to Kemausuor and Ackom (2016), this mechanism is designed to encourage consumers to shift their usage to off-peak times, thereby reducing strain on the grid. South Africa has implemented ToU pricing for some industrial and commercial consumers, encouraging them to use electricity during off-peak hours. ToU pricing requires advanced metering infrastructure, which is not widely available in many parts of SSA. Additionally, consumer awareness and ability to shift demand may be limited, reducing the effectiveness of this mechanism.
6. Block Tariffs:
   Block tariffs involve charging different rates based on the level of consumption. Typically, lower consumption blocks are charged at a lower rate, while higher consumption blocks face progressively higher tariffs. Olatokunbo and Iriobe (2022) suggested that this structure is designed to protect low-income consumers while encouraging energy efficiency among higher users. In countries like Tanzania and Zambia, block tariffs are used to make electricity more affordable for low-income households while ensuring that larger consumers pay rates that reflect their higher demand. Block tariffs can be difficult to implement effectively, particularly in areas with widespread electricity theft or inaccurate metering. Additionally, they may not always incentivize energy efficiency if the price difference between blocks is too small.
7. Market-Based Pricing:
   According to Malgas and Eberhard (2011), electricity prices can be determined by market forces, with prices fluctuating based on supply and demand in more liberalized markets. This is more common in wholesale markets where electricity is traded between generators and large consumers or distributors. The Southern African Power Pool (SAPP) operates a competitive electricity market where prices are determined through auctions and bilateral contracts between market participants. Market-based pricing can lead to volatility, especially in regions with

supply constraints or reliance on expensive generation sources. It also requires a well-developed infrastructure and regulatory environment to function effectively.

8. Pay-As-You-Go (PAYG) Models:
PAYG pricing is commonly used in off-grid solar markets, where consumers purchase electricity in small increments using mobile payments. This model makes electricity more accessible to low-income households who cannot afford the upfront costs of solar systems or grid connections IRENA (2020). Companies like M-KOPA in Kenya and BBOXX in Rwanda have successfully implemented PAYG models, allowing consumers to pay for solar energy on a daily or weekly basis. PAYG models have revolutionized access to electricity in rural areas, but they also face challenges such as ensuring payment reliability and maintaining affordability as system costs are recouped. The pricing mechanisms in SSA's energy markets are diverse, reflecting the region's varied economic conditions, regulatory environments, and energy access challenges. While traditional cost-plus and regulated tariffs remain prevalent, there is a growing shift towards more dynamic and market-based pricing models, particularly in regions with more developed infrastructure and regulatory frameworks. However, challenges such as affordability, infrastructure limitations, and regulatory capacity continue to impact the effectiveness and sustainability of these pricing mechanisms. Addressing these challenges will be key to ensuring that energy pricing in SSA supports the broader goals of expanding access, improving reliability, and encouraging investment in the energy sector.

4.2.5 Market Challenges and Trends

The energy market in SSA is characterized by a complex interplay of challenges and emerging trends that shape the region's ability to provide reliable and affordable energy to its growing population. A significant challenge is the region's severe infrastructure deficit. Many countries suffer from outdated and insufficient power generation facilities, inadequate transmission networks, and limited distribution infrastructure, especially in rural areas. These deficiencies result in frequent power outages, high technical losses, and a significant gap in electricity access, with nearly 775 million people in the region lacking access to electricity as of 2020. The financial instability of state-owned utilities, which dominate the market in many countries, exacerbates these issues.

Selassie and Hakobyan (2021) argued that these utilities often face significant financial losses due to inefficiencies, non-payment of bills, and tariff structures that do not cover the cost of service, leading to underinvestment in critical infrastructure. Another critical challenge is the regulatory and policy environment, which varies widely across the region. Eberhard and Godinho (2017) argued that in many countries, regulatory frameworks are underdeveloped or inconsistently enforced, creating an uncertain environment for investors and hindering private sector participation. This is particularly problematic for the renewable energy sector, where clear and supportive policies are essential for attracting investment. Political instability and governance issues further complicate the regulatory landscape, leading to delays in project implementation and increased risks for investors. Despite these challenges, there are several emerging trends that offer hope for the future of the energy sector in SSA. One of the most significant trends is the increasing penetration of renewable energy. Countries like Kenya, South Africa, and Ethiopia are leading the way in integrating renewable energy sources, particularly wind, solar, and geothermal, into their energy mix. This shift is driven by declining costs of renewable technologies, growing environmental awareness, and the need to diversify energy sources to enhance energy security. The rise of off-grid and mini-grid solutions, particularly in rural areas, is another important trend. These decentralized systems, often based on solar power, provide a viable alternative to traditional grid expansion and have the potential to significantly improve energy access in remote areas (Odhiambo 2021). Moreover, there is a growing emphasis on regional cooperation and cross-border electricity trade, facilitated by regional power pools such as the Southern African Power Pool (SAPP) and the West African Power Pool (WAPP). These initiatives aim to optimize the use of energy resources across borders, improve grid stability, and reduce the cost of electricity. However, the success of these initiatives depends on overcoming challenges related to infrastructure, regulatory harmonization, and political stability. To address these challenges and capitalize on emerging trends, there is a need for significant investment in infrastructure, the development of clear and supportive regulatory frameworks, and a focus on regional cooperation. Ensuring financial sustainability in the energy sector, particularly by reforming tariff structures and improving the efficiency of utilities, is also crucial. By addressing these issues, SSA can build a more resilient, sustainable, and inclusive energy market that supports broader economic and social development goals.

4.2.6 Energy Trading Platforms

The energy market in SSA is characterized by a complex interplay of challenges and emerging trends that shape the region's ability to provide reliable and affordable energy to its growing population. A significant challenge is the region's severe infrastructure deficit. Adewole et al. (2023) reported that many countries suffer from outdated and insufficient power generation facilities, inadequate transmission networks, and limited distribution infrastructure, especially in rural areas. These deficiencies result in frequent power outages, high technical losses, and a significant gap in electricity access, with nearly 775 million people in the region lacking access to electricity as of 2020.

The financial instability of state-owned utilities, which dominate the market in many countries, exacerbates these issues. These utilities often face significant financial losses due to inefficiencies, non-payment of bills, and tariff structures that do not cover the cost of service, leading to underinvestment in critical infrastructure. Another critical challenge is the regulatory and policy environment, which varies widely across the region. In many countries, regulatory frameworks are underdeveloped or inconsistently enforced, creating an uncertain environment for investors and hindering private sector participation. This is particularly problematic for the renewable energy sector, where clear and supportive policies are essential for attracting investment. Political instability and governance issues further complicate the regulatory landscape, leading to delays in project implementation and increased risks for investors. Despite these challenges, there are several emerging trends that offer hope for the future of the energy sector in SSA. One of the most significant trends is the increasing penetration of renewable energy. Countries like Kenya, South Africa, and Ethiopia are leading the way in integrating renewable energy sources, particularly wind, solar, and geothermal, into their energy mix. This shift is driven by declining costs of renewable technologies, growing environmental awareness, and the need to diversify energy sources to enhance energy security.

According to Duran and Sahinyazan (2021), the rise of off-grid and mini-grid solutions, particularly in rural areas, is another important trend. These decentralized systems, often based on solar power, provide a viable alternative to traditional grid expansion and have the potential to significantly improve energy access in remote areas. Moreover, there is a growing emphasis on regional cooperation and cross-border electricity trade, facilitated by regional power pools such as the Southern African Power Pool (SAPP) and the West African Power Pool (WAPP). These initiatives aim to optimize the use of energy resources across borders, improve grid stability, and reduce the cost of electricity. However, the success of these initiatives depends on overcoming challenges related to infrastructure, regulatory harmonization, and political stability (GFDRR 2023).

To address these challenges and capitalize on emerging trends, there is a need for significant investment in infrastructure, the development of clear and supportive regulatory frameworks, and a focus on regional cooperation. Ensuring financial sustainability in the energy sector, particularly by reforming tariff structures and improving the efficiency of utilities, is also crucial. By addressing these issues, SSA can build a more resilient, sustainable, and inclusive energy market that supports broader economic and social development goals.

4.3 Regulatory Frameworks

Regulatory frameworks in SSA's energy sector are complex and vary significantly across the region, reflecting the diverse political, economic, and social landscapes of individual countries. These frameworks are designed to govern the generation, transmission, and distribution of electricity, with the goal of ensuring reliable, affordable, and sustainable energy access. In many countries, state-owned utilities still dominate the energy sector, with regulatory bodies playing a crucial role in setting tariffs, approving investments, and overseeing market operations (IMF 2021). Andriarisoa et al. (2023) suggested that the effectiveness of these regulatory frameworks is often hindered by challenges such as political interference, limited technical capacity, and inconsistent enforcement of regulations. Some countries, like South Africa and Kenya, have made strides in liberalizing their energy markets and fostering private sector participation, particularly in the renewable energy sector. Despite these advancements, regulatory environments in many parts of the region remain underdeveloped, leading to issues such as inadequate infrastructure, financial instability of utilities, and slow progress in expanding energy access. As countries across SSA continue to reform and strengthen their regulatory frameworks, there is a growing focus on creating more transparent, independent, and effective regulatory bodies that can support the region's energy needs and drive sustainable development.

4.3.1 State Monopoly

State monopoly has historically dominated the electricity sector in SSA, often justified by the need for centralized control of a critical resource, safeguarding nascent domestic industries, and ensuring the provision of public services. This approach is rooted in the broader goal of maintaining national sovereignty over essential infrastructure and resources. Additionally, it reflects the developmental vision of many post-independence governments, which sought to prioritize access and equity in energy distribution. However, state monopoly in the region is often associated with significant challenges, such as operational inefficiencies, financial mismanagement, underinvestment, and susceptibility to corruption. These issues have hindered the expansion of electricity access and the improvement of service quality. The lack of competition in such a structure further exacerbates inefficiencies, as incentives for innovation and performance improvement remain minimal. An analysis of this policy framework across SSA reveals varying degrees of success and failure. In many cases, the lack of robust institutional and governance frameworks has compounded these issues, leading to an urgent need for reform through liberalization, private sector participation, or public-private partnerships (PPPs). These reform measures aim to introduce competition, attract private investment, and drive the efficiency gains needed to meet growing energy demands in the region. The economic impacts and challenges of state monopolies in SSA are summarized below:

1. Contribution to Economic Performance:
   State monopoly in the energy sector can have a significant influence on national economic performance or Gross Domestic Products (GDP). For example, state-owned enterprises contributed approximately 8% to Nigeria's GDP in 2020. However, inefficiencies within the power sector impose substantial economic costs, with Nigeria losing an estimated 29 billion US dollar annually due to unreliable electricity and lost business opportunities. This highlights the paradox of state monopoly — while they are vital to national economies, their inefficiencies can severely undermine economic growth.
2. Service Quality Issues:
   Monopolistic control often results in poor service quality due to the absence

of competitive pressure. In Kenya, despite reforms, electricity service remains inconsistent, with rural areas particularly underserved. Limited access to reliable electricity in rural regions hinders economic development and lowers the quality of life.

3. Investment and Innovation Deficits:
   State monopoly frequently stifles innovation and deters investment in infrastructure. For instance, in Zimbabwe, the Zimbabwe Electricity Supply Authority (ZESA) monopoly has led to underinvestment, resulting in frequent power outages and deteriorating infrastructure. This lack of modernization perpetuates inefficiency and fails to meet the growing energy demands of the population.
4. Access Challenges:
   Electricity access remains a significant issue in countries dominated by state monopolies. In Malawi, the Electricity Supply Corporation of Malawi (ESCOM) holds a monopoly, yet only about 13% of the rural population had access to electricity as of 2021. These access deficits reinforce socio-economic inequalities and hinder rural development.

4.3.2 Liberalization and Privatization

Liberalization and privatization in SSA emerged as part of broader economic reforms in the late 20th century (World Bank/AFD 2010; Hall et al. 2013). These reforms were driven by the need to address the inefficiencies of state-led economies, which often faced issues such as stagnation, corruption, and poor service delivery. Influenced by the global shift towards market-oriented policies and the recommendations of international financial institutions like the IMF and World Bank, many countries in the region embarked on ambitious reform agendas. Economic liberalization implies reducing state control over the economy, removing trade barriers, and encouraging foreign investment. This is intended to foster competition, enhance market efficiency, and integrate SSA into the global economy. Sector-specific liberalization focused on opening up specific sectors such as telecommunications, energy, and transport. Therefore, the liberalization of the energy sector allows private operators to enter the market, breaking the state monopolies that previously dominated. On the other end, privatization implies transfer of ownership or selling state-owned enterprises to private investors. This is done through direct sales, public offerings, or management contracts. The goal is to improve operational efficiency and service delivery by leveraging private sector expertise and investment. However, a systematic review conducted by Bensch (2019) on the impact of market-based reforms (liberalization, privatization, private sector involvement and regulation) on electricity access suggests that there is no sufficient evidence to infer that market-based reforms can enhance electricity access in developing countries. A brief account of the application of privatization as well as its challenges and criticisms in selected SSA countries is provided below:

1. South Africa:
   The adoption of the Electricity Regulation Amendment (ERA) Bill by the South African National Assembly in May 2024 marks a significant milestone in advancing the privatization of the electricity sector. This bill, among other provisions, establishes the framework for unbundling the national power utility (Eskom) — separating the transmission division from the rest of the power utility (Eskom), and to develop a multi-market electricity industry characterized by transparency, fairness, and modernization, which will contribute to a robust and efficient energy sector. The Bill will create an enabling environment for multiple electricity producers, such as IPPs, to compete on an equal footing. This competition is expected to drive down electricity prices, improve the reliability of supply, provide consumers with more options, attract increased investment in the electricity sector, and foster greater innovation and efficiency. Phalatse (2020) provided a historical background to the issues faced by the South African power utility until recently.
2. Nigeria:
   According to the International Trade Administration (2023), the Federal Republic of Nigeria initiated the privatization of its electricity sector in 2013. As a result, eleven electricity distribution companies (DisCos) and six generation companies (GenCos) were privatized to enhance efficiency and attract private sector investment. However, the transmission segment remained entirely under government ownership, with the Transmission Company of Nigeria (TCN) retaining 100% state control. This approach reflects a deliberate effort to maintain oversight of critical transmission infrastructure while encouraging private participation in other areas of the electricity value chain. Ahmed and Landi (2013) identified inadequate electricity infrastructure as the primary factor behind the persistent unreliability of power supply in Nigeria. They further contend that achieving the Nigerian government's goal of meeting electricity demand is contingent on substantial private sector investment.
3. Kenya:
   Karekezi and Mutiso (2000) highlighted significant milestones in Kenya's energy reform, including the adoption of the Electric Power Act of 1997 and the National Energy Policy of 2004. These legislative actions led to the establishment of the Rural Electrification Authority and the rationalization and unbundling of the state-owned electricity utility. The implementation of the 2010 Energy Policy further advanced privatization efforts, with KenGen partially privatized by offering 30% of its equity to private investors. Kenya's model is noteworthy for enabling private sector participation, both domestic and international, across all levels of the electricity value chain. This inclusive approach has positioned Kenya as the African leader in attracting IPPs.
4. DR Congo:
   According to the World Bank (2020), while legislation for liberalizing the electricity sector was introduced in 2014, its implementation was undermined by the absence of an effective and robust regulatory framework. The report further identifies weak governance, inadequate public financial capacity, and unfavorable policy environments as key barriers to scaling up decentralized renewable

energy solutions, despite the creation of the regulatory (ARE) and rural electification agencies (ANSER). These challenges hinder the development of sustainable energy systems, particularly in underserved regions, and underscore the importance of creating an enabling policy and institutional landscape for successful energy sector reform.

5. Challenges and Criticisms:
   a. Inequality and Access: privatization and liberalization have sometimes led to increased inequality, with rural and low-income areas often receiving less attention (Victor et al. 2015). Access to essential services like electricity has remained uneven in some cases, particularly in less economically developed regions.
   b. Regulatory and Governance Issues: effective regulation is crucial to ensure fair competition and protect consumers. In many cases, weak regulatory frameworks and governance challenges have hindered the success of privatization efforts.
   c. Economic Displacement: The transition to privatization has sometimes resulted in job losses and social displacement, as state-owned enterprises are restructured or downsized.

4.3.3 Public Private Partnership

Public-Private Partnerships (PPPs) in SSA play a significant role in addressing infrastructure building and electricity service delivery challenges. These partnerships are collaborations between governments and private sector entities to finance, build, and operate projects in energy infrastructure and systems. The goal is to leverage technical and financial strengths of both sectors to deliver energy services more efficiently. A few account of PPP cases in selected SSA countries are provided below:

1. South Africa:
   In 2023, the Presidential Climate Commission recommended adopting alternative investment strategies, including PPPs, to accelerate and scale up the expansion and strengthening of South Africa's electricity grid. The commission highlighted the importance of engaging the private sector to contribute to the financing, construction, and operation of newly expanded grid infrastructure. This initiative is set to operate under the framework of the National Transmission Company, which is being established as a critical component of the broader plan to unbundle Eskom into distinct entities for generation, transmission, and distribution. These measures aim to modernize the grid, enhance efficiency, and address South Africa's pressing energy challenges. In early 2024, the National Treasury announced plans to release a request for proposals (RFP) by the end of July for a pilot project aimed at leveraging off-balance-sheet financing to fast-track private-sector investment in transmission infrastructure. This initiative is designed to minimize the financial burden on Eskom's balance sheet and the national fiscus. The program

seeks to enable direct private-sector participation in the financing, construction, maintenance, and operation of grid infrastructure, paving the way for improved efficiency and expanded capacity in South Africa's energy transmission system (Bezuidenhoudt 2024).

2. Ghana:
   Sakyi (2019) highlighted the PPP established between the Electricity Distribution of Ghana (EDG) and the Manila Electric Company (MEC), a private equity firm. This partnership allocated 49% equity shares to the private entity, with the Ghanaian government retaining a controlling 51% stake. The partnership came in response to the shortcomings of Ghana's National Electrification Scheme, which struggled to achieve substantial rural electricity access through grid extension. Following an agreement with the World Bank in 2007, the Ghana Energy Development and Access Project (GEDAP) was initiated to advance electrification efforts, particularly through off-grid solutions such as solar photovoltaic (PV) systems funded by the Global Environment Facility (GEF). This approach not only facilitated rural electrification but also spurred the growth of solar merchants and rural banking institutions, creating a more robust ecosystem for renewable energy development (GPOBA 2018).
3. Uganda:
   Subsequent to unbundling in Uganda's electricity sector, which was realized in 2001, paving the way for privatization through concession agreements across the generation, transmission, and distribution segments. Under this framework, Eskom Uganda entered into a 20-year concession agreement to operate a hydroelectric power plant. Similarly, UMEME — a joint venture between Eskom Uganda (holding 44%) and Globeleq (holding 56%) — secured a 20-year concession to manage the country's electricity distribution network. These agreements marked significant steps in transitioning from a state-dominated model to a privatized framework aimed at improving efficiency, service delivery, and investment in Uganda's power sector.
4. DR Congo:
   The private sector, predominantly mining companies, has contributed significantly to expanding the DRC's electricity grid, although it has largely avoided traditional investment frameworks like IPPs and Build-Operate-Transfer (BOT) models. Instead, mining firms have primarily supported grid capacity growth by financing the rehabilitation and construction of electricity infrastructure through loans to SNEL. These investments have been driven by the sector's pressing need for reliable power to sustain operations. Notable projects include the refurbishment of the Inga II turbines, the construction of the Nzilo II Hydropower Plant (HPP), and enhancements to the Inga-Kolwezi high-voltage line. These efforts were funded by a 456 million USD credit extended by Kamoto Copper Company (KCC) and Mutanda Mining, subsidiaries of Glencore. Collectively, mining companies have provided over 1 billion USD in loans to SNEL, with investments primarily aimed at improving electricity supply for their operations. In addition to direct financing, mining firms have engaged in PPPs with local governments, such as the creation of Enerkac, where the Kasai Central provincial government

holds a 5% stake. Some companies have also opted to construct their own power plants to meet their energy demands, such as SOKIMO's 11 MW Budana HPP, which not only powers its mining activities but also supplies surplus electricity to nearby cities like Bunia and Mongbwalu. These initiatives highlight the strategic role of mining companies in the DRC's energy sector. In recent years, private sector-owned and operated isolated grids have emerged as a key development in expanding electricity access in the DRC. One example is Virunga SARL, a company established by the Virunga National Park's Virunga Foundation, which leveraged donor funding to develop and operate mini-grids. These include two hydropower plants (HPPs) with capacities of 0.38 MW and 1.35 MW in Mutwanga, as well as a 13.1 MW HPP at Matebe, currently serving 5,520 customers. Another notable operator, Electricité du Congo (EDC), rehabilitated the 1.6 MW Lugundi hydro plant, which now supplies electricity to 400 clients in Tshikapa. Similarly, Energie du Nord Kivu (ENK), a public-private partnership (PPP) between the Belgian company STS and the Nord Kivu provincial government, operates a grid in Butembo powered by a 1.8 MW HPP, serving 1,500 residential customers. Additionally, companies like Proton have secured concessions to operate grids but are still in the process of obtaining necessary funding to commence operations. These initiatives demonstrate the growing role of independent and PPP-operated grids in addressing energy challenges in isolated regions of the DRC (World Bank 2020).

5. Challenges of PPPs in SSA

   a. Regulatory and Institutional Frameworks: many SSA countries lack the necessary regulatory and institutional frameworks to support effective PPPs.
   b. Financial Constraints: securing financing for PPP projects can be challenging due to high perceived risks.
   c. Capacity and Expertise: there is often a lack of expertise and capacity within governments to manage PPP projects effectively.
   d. Political Risks: political instability and changes in government policies can affect the continuity and success of PPP projects.

6. Opportunities

   a. Renewable Energy: With the push for sustainable development, there is significant potential for PPPs in renewable energy projects, particularly in solar and wind energy.
   b. Urbanization: as urbanization accelerates, PPPs can play a critical role in developing infrastructure to support growing cities.
   c. Regional Integration: cross-border PPPs can help develop infrastructure that enhances regional integration, such as transport corridors and power pools.

When well structured and managed, PPPs constitute powerful mechanisms for addressing the infrastructure deficit and improving service delivery in SSA. However, they require strong governance, clear legal frameworks, and a commitment to transparency and accountability.

4.3.4 Power Purchase Agreement

A Power Purchase Agreement (PPA) in SSA serves as a foundational tool for the financing and development of energy projects, particularly within the renewable energy sector, which is critical for the region's sustainable growth. A PPA is a long-term contract — usually spanning 15 to 25 years — between an electricity producer, often a private sector entity such as an IPP, and a purchaser, typically a government-owned utility or a large corporate buyer. This contract stipulates the terms under which the electricity will be sold, including the price, quantity, and delivery schedule. The fixed price agreed upon in a PPA provides a predictable revenue stream for the electricity producer, which is crucial for securing financing from investors and lenders who require assurance of steady returns before committing capital to energy projects. In SSA, where the energy infrastructure is often underdeveloped, and access to electricity remains limited, PPAs have been instrumental in driving the adoption and expansion of renewable energy sources. For example, South Africa's Renewable Energy Independent Power Producer Procurement Programme (REIPPPP) is a flagship initiative that has leveraged PPAs to attract over 25 billion USD in investment since its launch in 2011. This program has led to the successful commissioning of more than 6,000 MW of renewable energy capacity, including solar, wind, and hydropower projects, making South Africa one of the leading renewable energy markets in the region. Another notable example is Kenya's Lake Turkana Wind Power project, one of the largest wind farms on the African continent with a capacity of 310 MW. This project was made possible through a PPA signed with the Kenya Power and Lighting Company (KPLC), which committed to purchasing the electricity generated by the wind farm at an agreed price. The Lake Turkana Wind Power project alone contributes approximately 15% of Kenya's total installed electricity capacity, significantly bolstering the country's energy mix and reducing its reliance on fossil fuels. However, despite the successes, the PPA landscape in SSA is fraught with challenges. One of the primary issues is currency risk, as PPAs are often denominated in foreign currencies such as the USD, while the utility's revenue is in the local currency. This mismatch can lead to financial instability if the local currency depreciates significantly against the foreign currency, increasing the cost of electricity purchases and putting strain on the utility's finances (Pillot et al. 2019). Furthermore, the creditworthiness of utilities in SSA is often a concern, as many are financially weak and struggle to meet their payment obligations under the terms of the PPA (Nuru et al. 2021; Bukari et al. 2021).

This uncertainty can deter investors and lenders, who may perceive the risk of default as too high. To address these risks, many PPA projects in SSA are supported by additional financial instruments, such as guarantees from multilateral development banks or partial risk guarantees from institutions like the World Bank. These guarantees provide an added layer of security for investors, ensuring that they will be compensated in the event of non-payment by the utility or other unforeseen challenges. This risk mitigation is essential for attracting the necessary capital to fund large-scale energy projects in the region. Despite these challenges, PPAs remain a vital mechanism for expanding energy access across SSA, where more than 600 million people still live without electricity. By providing the financial certainty needed to mobilize investment, PPAs enable the development of energy infrastructure that is essential for economic growth, job creation, and the achievement of sustainability goals. As SSA continues to develop its energy sector, PPAs will likely play an even more critical role in facilitating the transition to a more resilient, sustainable, and inclusive energy future.

4.3.5 Cross-Region Energy Trading

Cross-region energy trading in SSA is becoming increasingly significant as countries in the region seek to improve energy access, enhance grid stability, and optimize the use of renewable energy resources. This form of trading involves the exchange of electricity between regions or SSA pools, typically facilitated through interconnected power grids. The primary goal is to leverage the diverse energy resources across SSA to meet varying demand and supply conditions, reduce energy costs, and ensure a more reliable and sustainable power supply.

1. Importance of Cross-Region Energy Trading in SSA
   a. Resource Optimization: SSA is characterized by a wide variation in energy resources. For example, while Ethiopia has abundant hydropower potential, countries like South Africa and Botswana are rich in coal, and others like Kenya and Morocco have significant wind and solar resources. Cross-region energy trading allows these countries to share excess energy generated from their respective resources, ensuring that power is available where it is most needed.
   b. Enhancing Energy Security: energy security is a major concern in SSA, where many countries face frequent power shortages due to limited generation capacity, unreliable grid infrastructure, or dependency on a single energy source. By engaging in cross-border energy trading, countries can stabilize their energy supply, reduce their vulnerability to local disruptions, and diversify their energy mix.
   c. Economic Benefits: cross-region energy trading can lead to significant economic benefits. Countries with surplus power can export electricity to their neighbors, generating revenue that can be reinvested in domestic energy infrastructure. At the same time, importing countries can benefit from lower energy

costs, as they can purchase electricity at a lower price compared to the cost of local generation.

d. Facilitating Renewable Energy Integration: SSA is increasingly investing in renewable energy projects, but these sources are often intermittent and geographically dispersed. Cross-region energy trading can help balance the grid by sharing renewable energy across borders, ensuring that excess power generated from wind, solar, or hydropower in one country can be used to meet demand in another, thus reducing reliance on fossil fuels and lowering carbon emissions.

2. Challenges and Future Prospects While cross-region energy trading holds significant promise for SSA, several challenges remain:

    a. Infrastructure Gaps: The region's power infrastructure is often underdeveloped, with limited interconnections between countries, making cross-border energy trading difficult.
    b. Regulatory and Political Barriers: Differing regulatory frameworks, political instability, and lack of coordinated energy policies can hinder cross-region energy cooperation.
    c. Financial Constraints: Funding for the development and maintenance of cross-border transmission lines and other necessary infrastructure is often lacking.

    Despite these challenges, the future of cross-region energy trading in SSA is promising. Continued investment in power infrastructure, regional cooperation, and the integration of renewable energy sources are expected to drive the growth of energy trading in the region. As SSA countries work together to develop and harmonize their energy markets, cross-region energy trading will play a crucial role in achieving sustainable and inclusive economic development across the continent.

4.4 Conclusion

The transformation of energy markets and regulatory frameworks in SSA is crucial for the region's goals of expanding energy access, strengthening economic resilience, and achieving environmental sustainability. In recent years, numerous SSA countries have initiated significant reforms aimed at liberalizing their energy sectors. These reforms have been designed to reduce the dominance of state-owned utilities, open up markets to private sector participation, and attract much-needed investment, particularly in renewable energy projects. By encouraging competition and creating a more favourable environment for independent power producers (IPPs), these reforms have paved the way for increased investment in solar, wind, hydro, and geothermal energy, helping to diversify the energy mix and reduce reliance on fossil fuels.

One of the most notable developments in this context has been the establishment of regional power pools, such as the Southern African Power Pool (SAPP) and the

West African Power Pool (WAPP). These initiatives have been instrumental in facilitating cross-border energy trading, allowing countries to share surplus energy and balance supply and demand across the region. This has optimized the use of SSA's diverse energy resources, enhancing grid stability and reducing costs. For instance, hydropower-rich countries can export electricity during high rainfall periods, while countries with abundant solar or wind resources can do the same during peak generation times. This collaborative approach not only maximizes resource utilization but also strengthens energy security across the continent. However, despite these positive developments, SSA still faces considerable challenges in fully realizing the potential of its energy markets.

The region's energy infrastructure remains underdeveloped, with insufficient transmission and distribution networks that limit the efficient flow of electricity within and between countries. Regulatory fragmentation, where different countries have inconsistent or conflicting policies, further complicates cross-border cooperation and investment. Additionally, financial constraints, including limited access to capital and high costs of borrowing, continue to hinder the expansion and modernization of energy infrastructure. These challenges are exacerbated by political and economic instability in some areas, which can deter investors and slow the pace of reform. The integration of renewable energy into national grids presents additional challenges, requiring more advanced and flexible regulatory frameworks that can accommodate the intermittent nature of renewables and support the deployment of new technologies, such as energy storage and smart grids. These frameworks must also address issues related to grid stability, tariff structures, and the management of distributed energy resources. Without these advancements, the full benefits of renewable energy cannot be realized, and the transition to a low-carbon economy will be slowed down. Looking ahead, the success of SSA's energy markets will depend on the continuation and deepening of regulatory reforms that prioritize transparency, competition, and regional cooperation. These reforms should aim to create a more integrated and harmonized energy market across the region, with a focus on building robust infrastructure, improving regulatory alignment, and attracting sustainable financing. By doing so, SSA can develop resilient and sustainable energy systems capable of meeting the growing demands of its populations and supporting long-term economic development while also contributing to global efforts to combat climate change.

References

Adewole, A., Shipworth, M., Lemaire, X., Sanderson, D.: Peer-to-Peer energy trading, independence aspirations and financial benefits among Nigerian households. Energy Policy **174**, 113442 (2023). https://doi.org/10.1016/j.enpol.2023.113442

Ahmed, A., Landi, J.H.: Assessment of private sector financing of electricity infrastructure in Nigeria. J. Econ. Financ. **1**(5), 13–24 (2013)

Andriarisoa, M.N., Tambo, E.G., Tsuanyo, D., Nguedoung, A.N.: Peer-to-Peer Energy Trading Using Blockchain in Sub-Saharan Africa: Towards a Policy and Regulatory Framework, Practice, progress, and proficiency in sustainability, pp.123–159 (2023). https://doi.org/10.4018/978-1-7998-8638-9.ch006.

Bensch, G.: The effects of market-based reforms on access to electricity in developing countries: A systematic review. J. Dev. Effect. **11**(2), 165–188 (2019)

Bezuidenhoudt, L.: Financing South Africa's grid expansion on a Public-Private Partnership (PPP) basis a viable solution. Available at: https://www.fasken.com/en/knowledge/2024/06/financing-south-africas-grid-expansion-on-a-public-private-partnership-basis-a-viable-solution (2024). Last accessed Nov 2024

Blyden, B.K., Davidson, I.E.: Energizing Africa's emerging economy. IEEE Power Ener. Mag. **3**(4), 24–31 (2005). https://doi.org/10.1109/mpae.2005.1458227

Bukari, D., Kemausuor, F., Quansah, D.A., Adaramola, M.S.: Towards accelerating the deployment of decentralised renewable energy mini-grids in Ghana: Review and analysis of barriers. Renew. Sustain. Energy Rev. **135**, 110408 (2021). https://doi.org/10.1016/J.RSER.2020.110408

Copinschi, P.: Energy and the Economy in sub-saharan Africa. In: Hafner, M., Luciani, G. (eds.). In: The Palgrave Handbook of International Energy Economics. Palgrave Macmillan, Cham (2022). https://doi.org/10.1007/978-3-030-86884-0-34.

Dagnachew, A.G., Choi, S.M., Falchetta, G.: Energy planning in Sub-Saharan African countries needs to explicitly consider productive uses of electricity. Sci. Rep. **13**, 13007 (2023). https://doi.org/10.1038/s41598-023-40021-y

Duran, A.S., Sahinyazan, F.G.: An analysis of renewable mini-grid projects for rural electrification. Socio-Economic Planning Sciences, Elsevier, vol. 77(C) (2021). https://doi.org/10.1016/j.seps.2020.100999.

Eberhard, A., Godinho, C.: A review and exploration of the status, context and political economy of power sector the reform in Sub-Saharan Africa, South Asia, and Latin America. MIR Working Paper, University of Cape Town (2017)

Eberhard, A., Gratwick, K., Morella, E. and Antmann, P.: Independent Power Projects in Sub-Saharan Africa: Lessons from Five Key Countries. Directions in development,energy and mining Washington, D.C.: World Bank Group (2016). http://documents.worldbank.org/curated/en/795581467993175836

Global Facility for Disaster Reduction and Recovery (GFDRR).: Building regulations in Sub-Saharan Africa: A status review of the building regulatory environment. Available at: https://www.gfdrr.org/en/publication/building-regulations-sub-saharan-africa (2023). Accessed 11 Nov 2024

Global Parternership on Output-based Aid (GPOBA).: Expanding electrification to low-income households in rural Ghana with microfinance. Available at: http://documents1.worldbank.org/ (2018). Last accessed Nov 2024

Hall, D., Van Niekerk, S., Nguyen, J., Thomas, S.: Energy liberalisation, privatisation and public ownership. Available at: https://www.world-psi.org/sites/default/files/en-psiru-ppp-final (2013)

Igbinovia, F.O., Tlusty, I.J.T.: Electrical energy in Africa: The status of interconnections. World Eng. Conf. Sustain. Infrastruct. (WECSI 2014) at Abuja, Nigeria, pp. 1–14 (2014)

Ihirwe, J.P., Li, Z., Sun, K., Bimenyimana, S., Wang, C., Asemota, G.N.O., et al.: Solar PV minigrid technology: Peak shaving analysis in the East African community countries. Int. J. Photoenergy **2021**, 1–40 (2021). https://doi.org/10.1155/2021/5580264

International Energy Agency (IEA): Electricity Market Report–December (2020). https://www.iea.org/reports/electricity-market-report-december-2020

International Monetary Fund (IMF) Annual Report.: Resilience in the face of change. Available on: https://cdn.sanity.io/files/yg4ck731/production/305c76ee35a5e231ed9b2720ff64d8f980a4101e.pdf (2024). Last Accessed Nov 2024

International Monetary Fund (IMF).: The Global Informal Workforce Priorities for Inclusive Growth. Available at https://www.elibrary.imf.org/display/book/9781513575919/ch001.xmlhttps://doi.org/10.5089/9781513575919.071 (2021). Last accessed Nov 2024

International Renewable Energy Agency (IRENA).: PAY-AS-YOU-GO models innovation landscape brief. Available at: https://www.irena.org/-/media/Files/IRENA/Agency/Publication/2020/Jul/IRENA-Pay-as-you go-models-2020.pdf (2020)

International Trade Administration (ITA).: Electricity. Power Systems and Renewable Energy. Available at: https://www.trade.gov/country-commercial-guides/nigeria-electricity-and-power-systems (2023). Last accessed Nov 2024

IRENA.: Sub-Saharan Africa: Policies and finance for renewable energy deployment. In: International Renewable Energy Agency, Abu Dhabi (2024). https://www.irena.org/-/media/Files/IRENA/Agency/Publication/2024/Jul/IRENA-SS-Africa-policies-finance-RE-2024.pdf

IRENA and AfDB.: Renewable Energy Market Analysis: Africa and Its Regions. International Renewable Energy Agency and African Development Bank, Abu Dhabi and Abidjan. www.irena.org/publications (2022). Last accessed Nov 2024

Kambanda, C.: The African experience. In: Cottier, T., Espa, I. (eds.) In: International Trade in Sustainable Electricity: Regulatory Challenges in International Economic Law. Cambridge University Press, pp. 156–168 (2017)

Karekezi, S., Mutiso, D.: Power sector reform: A kenyan case study. In: Power Sector Reform in sub-Saharan Africa, pp. 83–120. Palgrave Mcmillan (2000)

Kemausuor, F. and Ackom, E.: Toward universal electrification in Ghana. Wiley Interdiscip. Rev.: Energy Environ. **6**(1), e225 (2016). https://doi.org/10.1002/wene.225.

Klug, T.W., Beyene, A.D., Meles, T.H., Toman, M.A., Hassen, S., Hou, M., Klooss, B., Mekonnen, A., Jeuland, M.: A review of impacts of electricity tariff reform in Africa. Energy Policy **170**, 113226 (2022). https://doi.org/10.1016/j.enpol.2022.113226

Luhangala, D.L., Ambole, A., Musango, J.K., Ceschin, F., Dulo, S.: Energy price modeling in sub-Saharan Africa: a systematic literature review. IOP Publish. **2**(1), 015001 (2022). https://doi.org/10.1088/2634-4505/ac3fee

Malgas, I., Eberhard, A.: Hybrid power markets in Africa: Generation planning, procurement and contracting challenges. Energy Policy **39**(6), 3191–3198 (2011). https://doi.org/10.1016/j.enpol.2011.03.004

Ndiritu, S.W., Engola, M.K.: The effectiveness of feed-in-tariff policy in promoting power generation from renewable energy in Kenya. Renew. Energy **161** (2020). https://doi.org/10.1016/j.renene.2020.07.082

Nuru, J.T., Rhoades, J.L., Gruber, J.S.: The socio-technical barriers and strategies for overcoming the barriers to deploying solar mini-grids in rural islands: Evidence from Ghana. Technol. Soc. **65**, 101586 (2021). https://doi.org/10.1016/J.TECHSOC.2021.101586

Odhiambo, N.M.: Trade openness and energy consumption in sub-Saharan African countries: A multivariate panel Granger causality test. Energy Rep. **7**, 7082–7089 (2021). https://doi.org/10.1016/j.egyr.2021.09.103

Olatokunbo, A.O., Iriobe, O.C.: Tariff pricing disparity and household electricity prepaid users' behavior in southwest, Nigeria. Christopher Univ. J. Manage. Social Sci. **2**(2), 95–104 (2022). Available at SSRN: https://ssrn.com/abstract=4219820 or http://dx.doi.org/10.2139/ssrn.4219820 Last Accessed Nov 2024

Organisation for Economic Co-operation and Development (OECD): Scaling Up Adaptation Finance in Developing Countries: Challenges and Opportunities for International Providers. Green Finance and Investment, OECD Publishing, Paris, (2023). https://doi.org/10.1787/b0878862-en

Oxford Institute for Energy Studies.: Issues in the pricing of domestic and internationally traded gas in MENA and sub-Saharan Africa. Available at: https://www.oxfordenergy.org/publications/issues-in-the-pricing-of-domestic-and-internationally-traded-gas-in-mena-and-sub-saharan-africa-2/ (2017). Accessed 11 Nov 2024

Phalatse, S.: Eskom: Historical background to its current crises. Inst. Econ. Justice Work. Paper Ser. **4** (2020)

Pillot, B., Muselli, M., Poggi, P., Dias, J.B.: Historical trends in global energy policy and renewable power system issues in Sub-Saharan Africa: The case of solar PV. Energy Policy **127**, 113–124 (2019). https://doi.org/10.1016/J.ENPOL.2018.11.049

Prasad, G.: Improving access to energy in sub-Saharan Africa. Curr. Opinion Environ. Sustain. **3**(4), 248–253 (2011). https://doi.org/10.1016/j.cosust.2011.05.003

Sakyi, K.A.: Public corporation monopolies-case study of sale of electricity company of Ghana. Adv. Soc. Sci. Res. J. **6**(4), 148–167 (2019)

Selassie, A.A., Hakobyan, S.: Six charts show the challenges faced by Sub-Saharan Africa. Available at: https://www.imf.org/en/News/Articles/2021/04/12/na041521-six-charts-show-the-challenges-faced-by-sub-saharan-africa (2021)

Streatfeild, J.: Investment in Africa's electricity sector: A historical overview and a way forward. J. Int. Commerce Econ. (2018). https://www.usitc.gov/journals

Sustainable Markets Initiative's Energy Transition Task Force.: Unlocking renewable energy development in sub-saharan Africa. Available at: https://www.sustainable-markets.org/news/public-private-sector-collaboration-vital-for-unlocking-renewable-energy-development-in-sub-saharan-africa-new-report-from-the-sustainable-markets-initiative-energy-transition-task-force/ (2022). Accessed 11 Nov 2024

United States Energy Information Administration (US EIA, 2020): Annual Energy Outlook 2020 with projections to 2050. https://www.eia.gov/outlooks/aeo/pdf/aeo2020%20full%20report.pdf

USAID. gov/Power Africa.: Hey Discos! Electricity distribution companies. Available at: https://medium.com/power-africa/hey-discos-that-is-electricity-distribution-companies-f0d124f562f4 (2018). Accessed 11 Nov 2024

Victor, O.E., Aziz, N.A., Jaffar, A.R.: Privatization of electricity service delivery in developing nations: issues and challenges. Int. J. Built Env. Sust. **2**(3), 202–210 (2015)

WHO, World Bank, IRENA and SE4All.: Energizing health: accelerating electricity access in health-care facilities (2023). https://www.irena.org/-/media/Files/IRENA/Agency/Publication/2023/Jan/IRENA-WHO-Energizing-health-care-2023.pdf

World Bank Group.: Off-Grid Solar Market Trends Report 2020 (2020). Available at: https://www.worldbank.org/en/topic/energy/publication/off-grid-solar-market-trends-report-2020

World Bank.: Increasing Access to Electricity in the Democratic Republic of Congo: Opportunities and Challenges. World Bank, Washington, DC. (2020). https://openknowledge.worldbank.org/handle/10986/33593

World Bank/AFD.: Africa's Infrastructure (2010). http://www.infrastructureafrica.org/

Chapter 5
Energy Access in Rural Communities: A Trade-Off Between Costs and Revenue

5.1 Introduction

Promoting access to energy is fundamental for driving economic development, fostering community progress, and achieving environmentally sustainable growth (Tazvinga et al. 2020). With more than 600 million people in SSA lacking access to electricity, addressing energy poverty in this region has become an urgent and critical priority. Zigah and Creti (2023) emphasize that improving access to energy services is more than just a developmental milestone — it is a pivotal step in alleviating poverty and enhancing the quality of life for millions.

Michoud and Hafner (2021) highlight that the reliance on solid fuels in these regions is deeply rooted in a combination of cultural practices and economic limitations, leading to significant repercussions for environmental sustainability, energy accessibility, and public health. Thorbecke (2013) and Monyei et al. (2022) argue that this widespread energy poverty constrains business opportunities, healthcare delivery, and educational progress, thereby obstructing social and economic development. Nevertheless, Bishoge et al. (2020) and Tiruye et al. (2021) underscore the untapped potential of the region's renewable energy resources — including hydroelectric, wind, and solar capacities — as a promising avenue to address these challenges. Harnessing these sustainable energy sources could act as a catalyst for transformative development and poverty reduction in the region.

Promoting energy access in SSA presents a multifaceted challenge requiring a comprehensive approach to resource allocation, infrastructure development, supportive policy frameworks, and addressing financial constraints. These factors collectively highlight the need for optimizing energy systems to ensure sustainable and equitable solutions (Falchetta et al. 2022). Therefore, multi-objective optimization (MOO) techniques could play a pivotal role in this process, enabling energy systems to be designed, operated, and expanded efficiently and cost-effectively (Suberu et al. 2013). The urgency of expanding energy access in SSA underscores the importance

P. N. Bokoro and G. Sharma, *Curbing Energy Poverty in Sub-Saharan Africa*,
https://doi.org/10.1007/978-3-031-90072-3_5

of providing not only wider electricity coverage but also reliable, clean, and affordable energy that supports long-term socio-economic development (Chirambo 2018; Jumbe and Mkondiwa 2013).

Achieving these outcomes demands a holistic strategy that integrates investments in infrastructure, innovative technologies and business models, and policies that encourage both public and private sector participation (Landa et al. 2020; Bazilian et al. 2012). Mathematical models and optimization algorithms are critical in formulating strategies for energy production, distribution, and consumption that align with the region's unique geographic, economic, and social conditions (Akanle et al., 2022). Furthermore, integrated approaches that consider technical, economic, and policy dimensions are essential for ensuring that energy access initiatives are both sustainable and inclusive (Abisuga-Oyekunle et al. 2019). This analysis highlights the potential of privatization, deregulation, and investment incentives in enhancing electricity availability, particularly in underserved rural areas. It also emphasizes the alignment of these measures with broader SDGs, such as eradicating poverty, advancing education, and fostering economic growth. By addressing these intertwined challenges, SSA can unlock its vast energy potential and drive progress across the region.

5.2 Economics of Rural Energy Systems

A lack of economic opportunities in rural areas of SSA is a significant barrier to attracting investment in centralized grid expansion and the development of decentralized energy systems. Low population densities, limited commercial activities, and low-income levels in these regions often lead to concerns about financial returns on infrastructure investments (Quansah et al. 2016). This discourages private and public stakeholders from committing resources to energy projects, as the demand for electricity may appear insufficient to justify high upfront costs and ongoing operational expenses.

To overcome this challenge, strategies such as integrating energy solutions with income-generating activities, leveraging renewable energy resources, and deploying innovative business models — like pay-as-you-go systems — are critical for improving economic viability and fostering investment in rural energy systems (Blimpo and Cosgrove-Davies 2019; Winklmaier et al. 2020; Kyriakarakos et al. 2020). In this study, crop farming is identified as a key rural economic activity capable of supporting investments in decentralized energy systems. RE-based off-grid systems are therefore designed, built, and operated to provide electricity to farming communities, enabling them to utilize motive power for essential agricultural processes such as water pumping, fodder chopping, threshing, grinding, and drying. In turn, financial returns generated from crops produced by these cooperatives are used to cover the investment and operational costs of the RE-based off-grid systems.

Therefore, attracting investments in rural energy systems, government agencies and private entities seeking capital for energy projects in SSA should prioritize integrated modelling frameworks. These frameworks should incorporate income projections from agricultural activities, energy generation potential, and infrastructure costs while being constrained by socio-environmental factors, regulatory frameworks, and policy options. By aligning energy supply with agricultural productivity, such models can highlight the economic feasibility and sustainability of rural energy investments.

This strategy demonstrates how energy systems not only fulfill local energy needs but also generate sufficient returns to cover operational and capital expenditures, making them appealing to potential investors. Moreover, focusing on the synergy between energy systems and agricultural output ensures that investments contribute to broader development objectives such as poverty reduction and food security, while meeting investor demands for financial viability and scalability.

5.3 Multi-criteria Decision-Making Framework

A Multi-Criteria Decision-Making (MCDM) framework is a structured methodology designed to assist in decision-making where multiple and often conflicting criteria need to be evaluated. The inherent conflict stems from the fact that enhancing one criterion or objective typically results in compromises in others. MCDM seeks to balance these trade-offs by systematically assessing alternatives based on their performance across the defined criteria. This approach facilitates the selection of the most suitable option by considering priorities, preferences, and constraints (Ehrgott 2005; Triantaphyllou 2000). This five-stage process begins with evaluating the economic viability of activities (crop farming) in rural areas, followed by a thorough assessment of renewable energy resources, which are critical for establishing off-grid energy systems. The third stage focuses on implementing regulatory policies to provide essential technical guidelines on reliability, tariffs, and other key factors. The fourth and central stage involves dynamic modeling and multi-objective optimization (MOO) aimed at achieving energy justice by addressing the three key dimensions of energy poverty: availability, accessibility, and affordability. Additionally, the framework incorporates a performance evaluation mechanism using the Multidimensional Energy Poverty Index (MEPI). The MCDM framework is visually summarized in the flowchart presented in Fig. 5.1.

5.3.1 Problem Conceptualization

For a rural community in SSA, this model proposes the establishment of farm holdings. Each cooperative consolidates households or small-scale farmers capable of cultivating similar crops. Access to energy services, such as electricity, in this community is expected to significantly enhance the productivity of each farm holding,

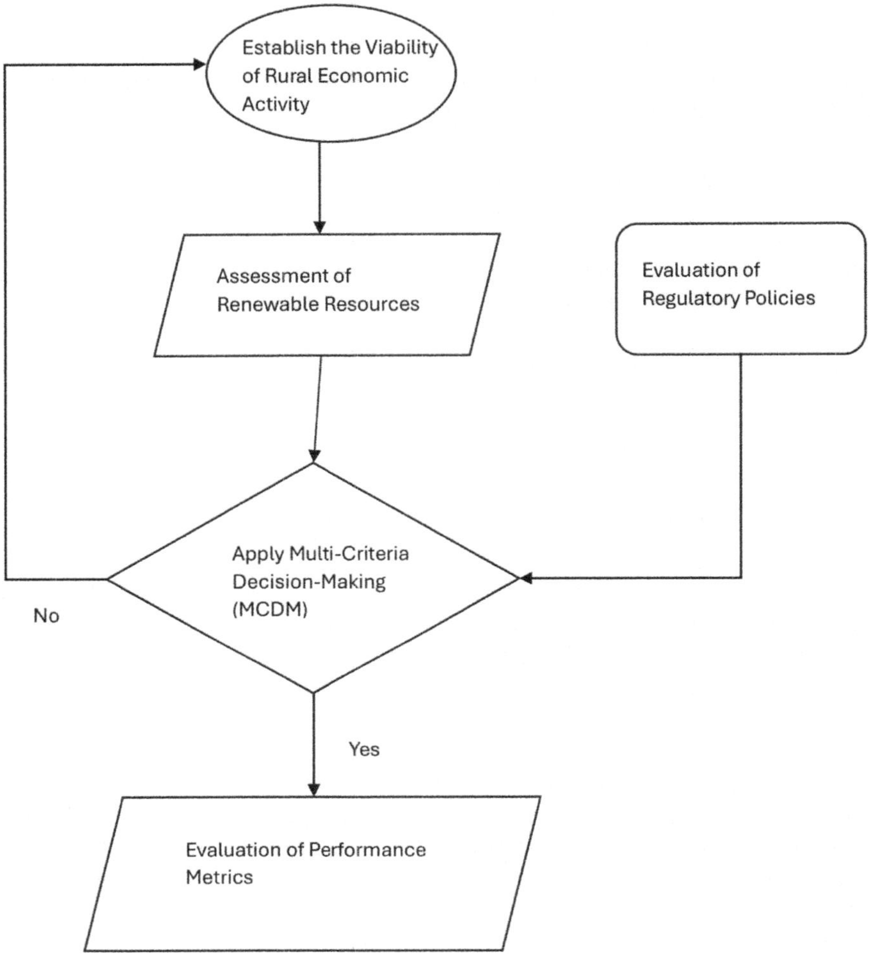

Fig. 5.1 MCDM framework

which will lead to increased crop yields, thereby boosting the income or revenue generated by farmers. However, these benefits come with associated costs, as farming entities must manage expenses related to acquiring energy services, labour and investing in advanced agricultural tools and technologies. Balancing these costs with the potential for increased income will be critical to ensuring the sustainability and profitability of farm holdings. Moreover, maintaining this balance is essential for ensuring the business model viability of RE-based off-grid energy system operators. These operators depend on increasing energy demand from farmers to generate sufficient profit, as farmers' ability to use electricity drives the long-term sustainability of the energy systems. By fostering an economically productive relationship, farm holdings benefit from enhanced agricultural output, while energy providers secure a reliable customer base, creating a mutually reinforcing cycle of growth and sustainability.

5.3.2 Decision Variables and Problem Formulation

Decision variables are adjustable parameters or factors within a system that directly impact the outcomes of decision criteria or objective functions. These variables are tailored to the context and objectives of the system under study, and could be categorized as either economical, technical, environmental, social or legal (Greco et al. 2016 ; Manos et al. 2013). For the purposes of this work, the following fundamental decision variables are adopted:

- Energy Consumption (E_k): this is the amount of electricity consumed by a farm holding k. It is expressed in kWh.
- Revenue (R_k): this consists of the revenue generated by a farm holding k from the sales of agricultural products.
- Land Area (A_k): this is the size of land in square-metres owned by a farm holding k.

The decision variables derived above form the basis for establishing key relationships between the energy company and the cooperative, which are essential for defining the objective functions. These relationships are presented as follows:

1. Total Income generated from sales of energy to a farming cooperative of n numbers of farm holdings: this represents the gross income made by the energy company by providing energy services to a cooperative consisting of n farmers. This could be expressed as follows:

$$\sum_{k=1}^{n} (p_e \cdot E_k + \alpha_k \cdot R_k)$$

   Where p_e is the price per unit of electricity sold to the farmers (set by the off-grid energy company), α_k is energy company's percentage share of each farmer's revenue.

2. Total Revenue generated by a cooperative: this is the sum of revenues generated by all farmers. This depends on the land size, the energy consumed, and the yield per unit area. It could be expressed by the following equation:

$$\sum_{k=1}^{n} R_k = \sum_{k=1}^{n} p_q \cdot Y_k \cdot A_k \tag{5.1}$$

   Where p_q is the price per unit of agricultural product sold by the farmers. Y_k is the quantity of yield produced per unit area by a farmer k.

3. Total Electricity Consumption of a cooperative: this represents the total electricity consumption by all farmers. The variation in land size cultivated by each farmer is a key determinant of electricity usage. Larger land holdings typically result in higher electricity demand due to increased irrigation, crop processing, and storage requirements. This relationship can be quantified mathematically as follows:

$$\sum_{k=1}^{n} E_k = \sum_{k=1}^{n} \beta_k \cdot A_k \tag{5.2}$$

Where β_k is the electricity consumption coefficient in kWh per unit of land area.

4. Total Costs incurred by the energy company (C_{tot}): this includes the total combined expenses associated with electricity generation, operational activities, and maintenance requirements. It could be noted as follows:

$$C_{tot} = C_{O\&M} + C_{Gen} \tag{5.3}$$

Where ($C_{O\&M}$) is aggregated operational and maintenance expenses. (C_{Gen}) is total expenses related to generation cost. These expenses could be expressed as follows:

$$C_{Gen} = \frac{C_{Cap}}{L \cdot E_{Gen}} \tag{5.4}$$

Where C_{Cap} is the capital investment, L is the plant lifespan, and E_{Gen} is the total kWh generated by the plant.

$$C_{O\&M} = \sum_{k=1}^{n} \frac{p_e \cdot E_k + \alpha_k \cdot R_k}{p_e \cdot E_k} \tag{5.5}$$

5. Land Growth of a cooperative: this could be modelled as follows:

$$\sum_{k=1}^{n} A_k^{t+1} = \sum_{k=1}^{n} A_k^t + \gamma_k \cdot [R_k - (p_e \cdot E_k + \alpha_k \cdot R_k)] \tag{5.6}$$

Where A_k^{t+1} is the land area of a farmer k in time $(t+1)$, A_k^t is the land area of a farmer k in time t, and γ_k the fractional remaining profit reinvested into land for a farm k.

5.3.3 Objective Functions

The dynamic interaction between the energy company and farming entities functions as a feedback loop: as crop farming entities expand their cultivated land, energy consumption increases, leading to higher electricity demand and, consequently, greater profits for the energy company. This relationship directly ties the company's revenue to key dimensions of energy poverty — accessibility, availability, and affordability — facilitating improved energy access for farmer holdings, which are essentially households engaged in farming activities. Therefore, the energy company's strategy should prioritize maximizing revenue while concurrently minimizing the total costs associated with operations, maintenance, and capital investment repayments. This could be expressed as follows:

1. Maximize the revenue (f_1):

$$Max.f_1 = \sum_{k=1}^{n} (p_e \cdot E_k + \alpha_k \cdot R_k) \tag{5.7}$$

   Where f_1 is the revenue objective function.

2. Minimize the costs (f_2):

$$Min.f_2 = \frac{C_{Cap}}{L \cdot E_{Gen}} + \sum_{k=1}^{n} \frac{p_e \cdot E_k + \alpha_k \cdot R_k}{p_e \cdot E_k} \tag{5.8}$$

   Where f_2 is the cost objective function.

The multi-objective function is given as:

$$\pi = \sum_{k=1}^{n} (p_e \cdot E_k + \alpha_k \cdot R_k) - \left(\frac{C_{Cap}}{L \cdot E_{Gen}} + \sum_{k=1}^{n} \frac{p_e \cdot E_k + \alpha_k \cdot R_k}{p_e \cdot E_k} \right) \tag{5.9}$$

5.3.4 Constraints

The optimization equation proposed in (5.9) is attained under the following set of constraints:

1. The total energy demand by a cooperative must be less or equal to the total energy generated by the supply company: this is expressed as follows:

$$\sum_{k=1}^{n} E_k \leq E_{Solar} + E_{Battery} + E_{Biomass}$$

2. The land growth constraint: the area land cultivated at time $(t + 1)$ must be greater or equal to that used at time t:

$$\sum_{k=1}^{n} A_k^{t+1} \geq \sum_{k=1}^{n} A_k^t$$

3. Non-Negativity constraint: energy consumed by a cooperative as well area land must not be negative in any given point:

$$\sum_{k=1}^{n} E_k \geq 0$$
$$\sum_{k=1}^{n} A_k \geq 0.$$

5.4 Non-dominated Sorting Genetic Algorithm II

The Non-Dominated Sorting Genetic Algorithm II (NSGA-II) consists of a robust evolutionary algorithm designed to solve multi-objective optimization (MOO) problems by identifying a set of Pareto-optimal solutions (Chaabani et al. 2023). This technique is employed to determine the Pareto optimal solutions (trade-offs) between costs and revenue for a case study described below.

5.4.1 Implementation of the NDSGA II

This algorithm is implemented in order to determine the trade-offs between cost and revenue for a 1.2 million USD 1 MW solar PV plant to power a rural community with farming entities. Farthing et al. (2023) and Birhanu et al. (2023) suggest that the monthly electricity consumption for a general crop farming is estimated to be between 900 and 2800 kWh. The World Bank Group (2017) and Giller et al. (2021) report that the monthly income of cropping farms is estimated to be between 500 to 1500 USD, and the electricity cost per kWh is 0.21 USD in most rural areas supplied from off-grid systems. The plant lifespan, which in this case could be regarded as the time to transfer plant ownership, is set to 10 years. The bounds of the decision variables (E_k) and (R_k) as well as the population size N are set in strict respect of the constraints. This is given is given as:

$$E_k \in [900, 2800]$$
$$R_k \in [500, 1500]$$
$$N = 22$$

For each individual i in the population, uniform random sampling is used for the decision variables. The following relationships are therefore observed:

$$E_k^i = E_k^{min} + 0.1 \cdot (E_k^{min}) \tag{5.10}$$

$$R_k^i = R_k^{min} + 0.05 \cdot (R_k^{min}) \tag{5.11}$$

The factors p_e and α_k as well as the total cost incurred by the energy company per farm are indicated in Table 5.1.

Table 5.1 Decision variables

E_k	R_k	p_e	α_k	C_{tot}
900	500	0.21	0.1	121.265
990	525	0.21	0.1	121.253
1080	550	0.21	0.1	121.243
1170	575	0.21	0.1	121.234
1260	600	0.21	0.11	121.159
1350	625	0.21	0.11	121.242
1440	650	0.21	0.11	121.236
1530	675	0.21	0.11	121.231
1620	700	0.21	0.12	121.247
1710	725	0.21	0.12	121.242
1800	750	0.21	0.12	121.238
1890	775	0.21	0.12	121.234
1980	800	0.21	0.13	121.25
2070	825	0.21	0.13	121.247
2160	850	0.21	0.13	121.244
2250	875	0.21	0.13	121.241
2340	900	0.21	0.14	121.256
2430	925	0.21	0.14	121.254
2520	950	0.21	0.14	121.251
2610	975	0.21	0.14	121.249
2700	1000	0.21	0.15	121.265
2790	1025	0.21	0.15	121.262
kWh	USD	USD/kWh	–	USD

5.4.2 *Pareto Optimality and Trade-Offs*

Using the data points assigned to the decision variables (refer to Table 5.1), the objective functions f_1 and f_2 are computed using Python programming language. The results are visualized through a Pareto Front (x-axis is the cost objective function and y-axis is the revenue objective function), which highlights the Pareto-optimal solutions for $\alpha_k \in [0.1, 0.11, 0.12, 0.13, 0.14, 0.15]$. The non-dominated solutions on the Pareto Front represent the trade-offs between the revenue generated and the costs incurred by the energy company, enabling a balanced decision-making process. This is shown in Figs. 5.2, 5.3, 5.4, 5.5, 5.6 and 5.7. Along the Pareto Front, it could be observed that an increase in revenue is typically accompanied by a corresponding decrease in cost and vice-versa. The points not located on the Pareto Front, such as those scattered below or farther away, represent dominated solutions. These are considered suboptimal because superior trade-offs exist — either achieving lower costs for the same revenue or higher revenue for the same cost. Higher revenues and reduced costs for the energy company are essential for enabling access to energy services. For crop farmers with monthly incomes ranging from 1000 to 1025 USD and $\alpha_k = 0.15$, the Pareto front identifies an optimal monthly revenue of 735.95

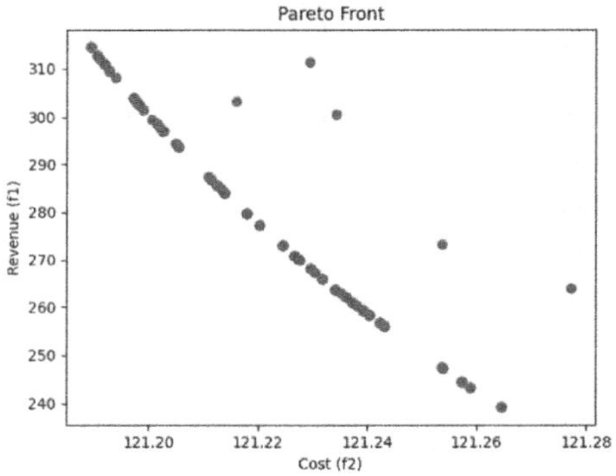

Fig. 5.2 Pareto front ($\alpha_k = 0.1$)

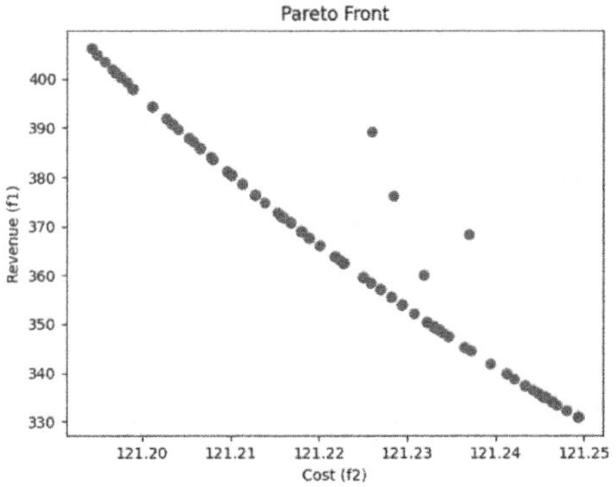

Fig. 5.3 Pareto front ($\alpha_k = 0.11$)

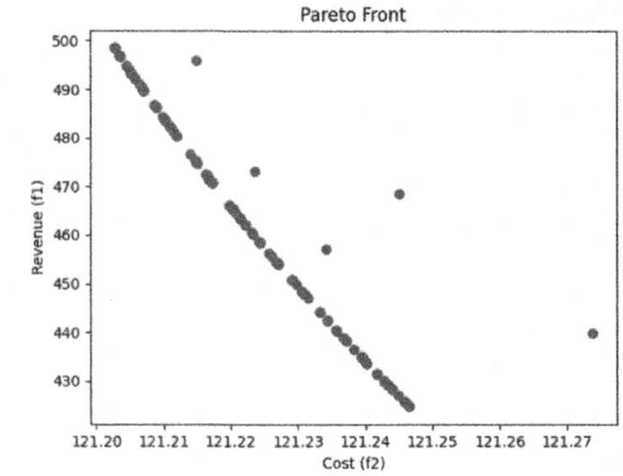

Fig. 5.4 Pareto front ($\alpha_k = 0.12$)

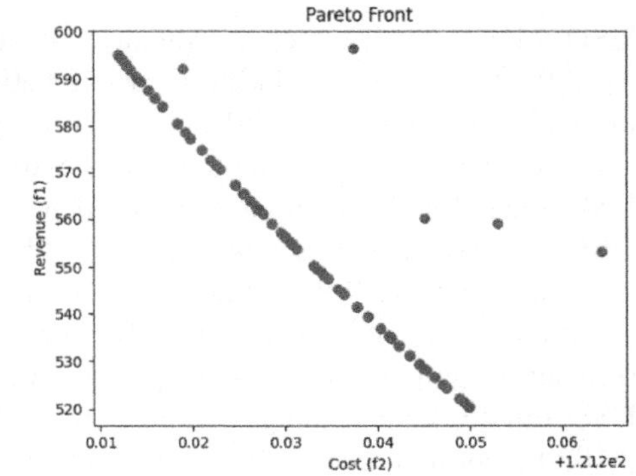

Fig. 5.5 Pareto front ($\alpha_k = 0.13$)

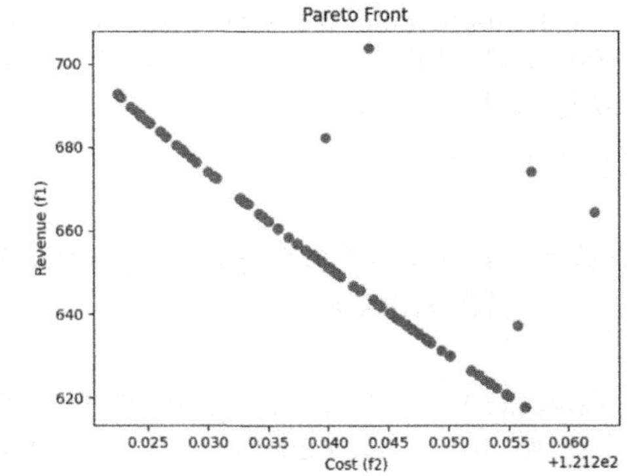

Fig. 5.6 Pareto front ($\alpha_k = 0.14$)

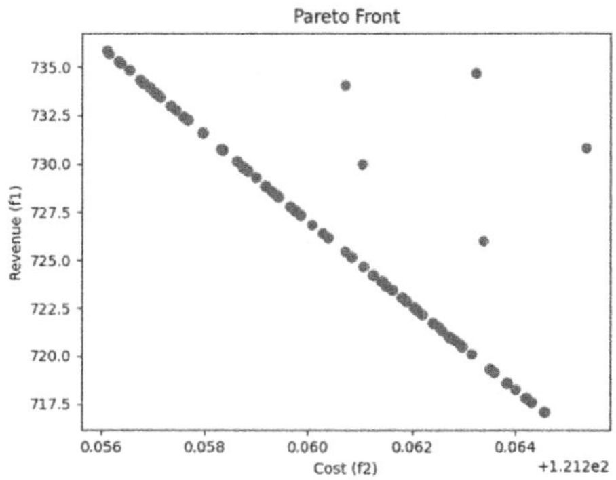

Fig. 5.7 Pareto front ($\alpha_k = 0.15$)

USD collected by the energy company, with an associated cost of 121.256 USD. For $\alpha_k = 0.14$, applicable to farmers earning between 900 and 975 USD per month, the optimal revenue is 692.7 USD, with a corresponding cost of 121.25613 USD.

When $\alpha_k = 0.13$, relevant to farmers with monthly incomes between 800 and 875 USD, the optimal revenue is 595.5 USD per month, incurred at a cost of 121.21179 USD. For farmers earning between 700 and 775 USD per month and $\alpha_k = 0.12$, the optimal revenue is 499.4 USD, with a cost of 121.2024 USD.

At $\alpha_k = 0.11$, for farmers with incomes between 600 and 675 USD per month, the energy company achieves an optimal revenue of 405.9 USD at a cost of 121.1946 USD. Finally, for consumers earning between 500 and 575 USD per month, with $\alpha_k = 0.10$, the optimal revenue is calculated to be 314.3 USD, with a cost of 121.18956 USD.

This implies that the annual revenue generated by the energy company will be 138034.32 USD for an annual cost of 31999.14792 USD, and annual net profit of 106035.1721 USD.

5.4.3 *Impact of MCDM Framework on Energy Poverty Alleviation*

According to the MEPI outlined in Chap. 2, energy poverty is identified when at least two out of five dimensions — availability, accessibility, affordability, energy consumption, and energy services — are not met. Given that rural areas in SSA are predominantly challenged by accessibility issues, the proposed MCDM framework is designed to address and enhance the dimensions of availability, accessibility, and affordability to alleviate energy poverty effectively.

1. Availability:
   The implementation of the MCDM framework depends on effective collaboration among funding institutions, regulatory bodies, energy investment companies,

government agencies or private sector entities, and rural communities. This coordinated approach promotes the sustainable and profitable development of renewable energy-based decentralized systems, delivering essential energy services to rural communities while supporting crop farming — a predominant and deeply rooted economic activity in rural SSA.

2. Accessibility:
   The implementation of the MCDM framework envisions electricity usage for small-scale farming activities located in the backyards of rural households. This arrangement inherently establishes an electrical connection between rural households and a decentralized off-grid system, operated by an energy company or supply authority, signifying an essential step toward achieving electricity access. Furthermore, the inclusion of electricity for household consumption contributes to the overall energy demand, thereby increasing the total revenue generated by the energy company.
3. Affordability:
   The implementation of the MCDM framework ensures a steady income, (R_k), for households generated from agricultural yields. This income serves as a crucial mechanism to cover the costs associated with energy services, shared between households and the energy company. A portion of the household income, represented as $(\alpha_k \cdot R_k)$, is incorporated into the total revenue accrued by the energy company from each consumer. This arrangement reinforces the financial sustainability or the affordability of energy service provision in rural areas.

This framework highlights that improving energy accessibility in rural areas of SSA requires a synergistic approach that integrates the three critical dimensions of energy poverty — availability, accessibility, and affordability. It underscores the limitations of a unidimensional strategy in addressing energy poverty, particularly in the rural regions of SSA.

5.5 Conclusion

Ineffective strategies for addressing energy poverty in rural SSA persist as significant obstacles, hindering efforts to alleviate this condition and mitigate its widespread impacts. Despite the establishment of numerous rural electrification agencies by governments across the continent, progress toward universal energy access remains limited, even with the backing of progressive regulatory policies and available funding. In many documented instances, the enduring lack of energy services in rural communities is predominantly perceived as an accessibility issue, often addressed from this narrow perspective.

This section advocates for a multi-dimensional approach to tackling energy poverty, asserting that the problem is best addressed through a trilogy of interconnected dimensions: availability, accessibility, and affordability. The proposed MCDM framework employs a multi-objective optimization strategy that balances the trade-offs between costs and revenues associated with energy services in rural areas. Costs are linked to ensuring energy availability, while revenues play a pivotal role in addressing accessibility and affordability dimensions.

Furthermore, the framework highlights the dynamic interplay between RE-based decentralized energy systems and rural households, where crop farming serves as a cornerstone of economic activity. This interdependence reinforces the critical relationship between energy access and community livelihoods. By fostering community empowerment and advancing SDGs, this comprehensive approach offers a pathway to transformative growth in rural SSA.

References

Abisuga-Oyekunle, O.A., Patra, S.K., Muchie, M.: SMEs in sustainable development: their role in poverty reduction and employment generation in sub-Saharan Africa. Afr. J. Sci. Technol. Innov. Dev. **12**(4), 405–419 (2019). https://doi.org/10.1080/20421338.2019.1656428

Akanle, O., Kayode, D., Abolade, I.: Sustainable development goals (SDGs) and remittances in Africa. Cogent Soc. Sci. **8**(1), 2037811 (2022). https://doi.org/10.1080/23311886.2022.2037811

Bazilian, M., Nussbaumer, P., Rogner, H., Brew-Hammond, A., Foster, V., Pachauri, S., Williams, E., Howells, M., Niyongabo, P., Musaba, L., Gallachóir, Ó.B., Radka, M., Kammen, D.M.: Energy access scenarios to 2030 for the power sector in sub-Saharan Africa. Util. Policy **20**(1), 1–16 (2012). https://doi.org/10.1016/j.jup.2011.11.002

Birhanu, B.Z., et al.: Solar-based irrigation systems as a game changer to improve agricultural practices in sub-Sahara Africa: a case study from Mali. Front. Sustain. Food Syst. Sec. Land Livelihoods Food Secur. **7** (2023). https://doi.org/10.3389/fsufs.2023.1085335.

Bishoge, O.K., Kombe, G.G., Mvile, B.N.: Renewable energy for sustainable development in sub-Saharan African countries: challenges and way forward. J. Renew. Sustain. Energy **12**(5), 052702 (2020). https://doi.org/10.1063/5.0009297

Blimpo, M.P., Cosgrove-Davies, M.: Electricity access in sub-Saharan Africa uptake, reliability, and complementary factors for economic impact (2019). https://openknowledge.worldbank.org/. Accessed Dec 2024

Chaabani, A., Karaja, M., Said, L.B.: An efficient non-dominated sorting genetic algorithm for multi-objective optimization. In: 9th International Conference on Control, Decision and Information Technologies (CoDIT), Rome, Italy, pp. 1565–1570 (2023). https://doi.org/10.1109/CoDIT58514.2023.10284357

Chirambo, D.: Towards the achievement of SDG 7 in sub-Saharan Africa: Creating synergies between Power Africa Sustainable Energy for All and climate finance in-order to achieve universal energy access before 2030. Renew. Sustain. Energy Rev. **94**, 600–608 (2018). https://doi.org/10.1016/j.rser.2018.06.025

Ehrgott, E.M.: Multicriteria Optimization, 2nd edn. Springer (2005). https://link.springer.com/book/10.1007/3-540-27659-9. Accessed Nov 2024

Falchetta, G., Adeleke, A., Awais, M., Byers, E., Copinschi, P., Duby, S., Hughes, A., Ireland, G., Riahi, K., Rukera-Tabaro, S., Semeria, F., Shendrikova, D., Stevanato, N., Troost, A., Tuninetti, M., Vinca, A., Zulu, A., Hafner, M.: A renewable energy-centred research agenda for planning and financing Nexus development objectives in rural sub-Saharan Africa. Energy Strat. Rev. **43** (2022). https://doi.org/10.1016/j.esr.2022.100922

Farthing, A., et al.: Quantifying agricultural productive use of energy load in Sub-Saharan Africa and its impact on microgrid configurations and costs. Appl. Energy **2023** (2023). https://doi.org/10.1016/j.apenergy.2023.121131

Giller, K.E., Delaune, T., Silva, J.V., et al.: Small farms and development in sub-Saharan Africa: Farming for food, for income or for lack of better options? Food Sec. **13**, 1431–1454 (2021). https://doi.org/10.1007/s12571-021-01209-0

Greco, S., Figueira, J., Ehrgott, M.: . Multiple Criteria Decision Analysis, vol. 37. Springer (2016)

Jumbe, C., Mkondiwa, M.: Comparative analysis of biofuels policy development in Sub-Saharan Africa: the place of private and public sectors. Renew. Energy **50**, 614–620 (2013). https://doi.org/10.1016/j.renene.2012.07.023

Kyriakarakos, G., Balafoutis, A.T., Bochtis, D.: Proposing a paradigm shift in rural electrification investments in sub-Saharan Africa through agriculture. Sustainability **12**(8), 3096 (2020). https://doi.org/10.3390/su12083096

Landa, T.F., Sales, E.A., Nascimento, A.N., Padula, A.D.: Optimizing the cooperated multi-countries biodiesel production and consumption in Sub-Saharan Africa. Energies **13**(18) (2020). https://doi.org/10.3390/en13184717

Manos, B., Chatzinikolaou, P., Kiomourtzi, F.: Sustainable Optimization of Agricultural Production. ICESD. January 19–20, Dubai, United Arab Emirates (2013)

Michoud, B., Hafner, M.: Energy Access in Sub-Saharan Africa: General Context in Financing Clean Energy Access in Sub-Saharan Africa: Risk Mitigation Strategies and Innovative Financing Structures. Springer International Publishing, vol. 2021, pp. 7–26 (2021)

Monyei, C.G., Akpeji, K.O., Oladeji, O., Babatunde, O.M., Aholu, O.C., Adegoke, D., Imafidon, J,O.: Regional cooperation for mitigating energy poverty in Sub-Saharan Africa: a context-based approach through the tripartite lenses of access, sufficiency, and mobility. Renew. Sustain. Energy Rev. **159**(C) (2022). (Elsevier)

Quansah, D.A., Adaramola, M.S., Mensah, L.D.: Solar photovoltaics in sub-Saharan Africa: addressing barriers, unlocking potentials. Energy Proc. (Elsevier) **106**, 97–110 (2016). https://doi.org/10.1016/j.egypro.2016.12.108

Suberu, M.Y., Mustafa, M.W., Bashir, N., Muhamad, N.A., Mokhtar, A.S.: Power sector renewable energy integration for expanding access to electricity in sub-Saharan Africa. Renew. Sustain. Energy Rev. **25**, 630–642 (2013). https://doi.org/10.1016/j.rser.2013.04.033

Tazvinga, H. Dzobo, O., Mapako, M.: Towards sustainable energy system options for improving energy access in Southern Africa. J. Energy S. Afr. **31**(2), 59–72 (2020). http://dx.doi.org/10.17159/2413-3051/2020/v31i2a6504

Thorbecke, E.: The interrelationship linking growth, inequality and poverty in Sub-Saharan Africa. J. Afr. Econ. **22**(suppl-1), i15–i48 (2013). https://doi.org/10.1093/jae/ejs028

Tiruye, G.A., Besha, A.T., Mekonnen, Y.S., Benti, N.E., Gebreslase, G.A., Tufa, R.A.: Opportunities and challenges of renewable energy production in Ethiopia. Sustainability **13**, 10381 (2021). https://doi.org/10.3390/su131810381.

Triantaphyllou, E.: Multi-criteria Decision Making Methods A Comparative Study (2000). https://link.springer.com/book/10.1007/978-1-4757-3157-6. Accessed Nov 2024

Winklmaier, J., Santos, S.A.B., Trenkle, T.: Economic development of rural communities in sub-Saharan Africa through decentralized energy-water-food systems. In: Regional Development in Africa (2020). https://www.intechopen.com/books. Accessed Dec 2024

World Bank Group: Income Diversification Patterns in Rural Sub-Saharan Africa (2017). https://documents.worldbank.org/en/publication/documents-reports/documentdetail/861761468340796327/income-diversification-patterns-in-rural-sub-saharan-africa-reassessing-the-evidence. Accessed Dec 2024

Zigah, E., Creti, A.: A comparative analysis of electricity access initiatives in Sub-Saharan Africa. In: Gromek-Broc, K. (eds) Regional Approaches to the Energy Transition. Springer, Cham (2023). https://doi.org/10.1007/978-3-031-19358-3-16.

The manufacturer's authorised representative in the EU is Springer
Nature Customer Service Centre GmbH, Europaplatz 3, 69115 Heidelberg,
Germany. If you have any concerns regarding our products, please
contact ProductSafety@springernature.com

Printed and bound by CPI Group (UK) Ltd, Croydon, CR0 4YY
30/09/2026
02225513-0003